Wolfgang Brezna

Scanning Capacitance Microscopy

Wolfgang Brezna

Scanning Capacitance Microscopy

and Spectroscopy on Semiconductor Materials

Südwestdeutscher Verlag für Hochschulschriften

Impressum/Imprint (nur für Deutschland/ only for Germany)
Bibliografische Information der Deutschen Nationalbibliothek: Die Deutsche Nationalbibliothek verzeichnet diese Publikation in der Deutschen Nationalbibliografie; detaillierte bibliografische Daten sind im Internet über http://dnb.d-nb.de abrufbar.
Alle in diesem Buch genannten Marken und Produktnamen unterliegen warenzeichen-, markenoder patentrechtlichem Schutz bzw. sind Warenzeichen oder eingetragene Warenzeichen der jeweiligen Inhaber. Die Wiedergabe von Marken, Produktnamen, Gebrauchsnamen, Handelsnamen, Warenbezeichnungen u.s.w. in diesem Werk berechtigt auch ohne besondere Kennzeichnung nicht zu der Annahme, dass solche Namen im Sinne der Warenzeichen- und Markenschutzgesetzgebung als frei zu betrachten wären und daher von jedermann benutzt werden dürften.

Verlag: Südwestdeutscher Verlag für Hochschulschriften Aktiengesellschaft & Co. KG
Dudweiler Landstr. 99, 66123 Saarbrücken, Deutschland
Telefon +49 681 37 20 271-1, Telefax +49 681 37 20 271-0, Email: info@svh-verlag.de
Zugl.: Vienna, Vienna University of Technology, Diss., 2005

Herstellung in Deutschland:
Schaltungsdienst Lange o.H.G., Berlin
Books on Demand GmbH, Norderstedt
Reha GmbH, Saarbrücken
Amazon Distribution GmbH, Leipzig
ISBN: 978-3-8381-0267-2

Imprint (only for USA, GB)
Bibliographic information published by the Deutsche Nationalbibliothek: The Deutsche Nationalbibliothek lists this publication in the Deutsche Nationalbibliografie; detailed bibliographic data are available in the Internet at http://dnb.d-nb.de.
Any brand names and product names mentioned in this book are subject to trademark, brand or patent protection and are trademarks or registered trademarks of their respective holders. The use of brand names, product names, common names, trade names, product descriptions etc. even without a particular marking in this works is in no way to be construed to mean that such names may be regarded as unrestricted in respect of trademark and brand protection legislation and could thus be used by anyone.

Publisher:
Südwestdeutscher Verlag für Hochschulschriften Aktiengesellschaft & Co. KG
Dudweiler Landstr. 99, 66123 Saarbrücken, Germany
Phone +49 681 37 20 271-1, Fax +49 681 37 20 271-0, Email: info@svh-verlag.de

Copyright © 2009 by the author and Südwestdeutscher Verlag für Hochschulschriften Aktiengesellschaft & Co. KG and licensors
All rights reserved. Saarbrücken 2009

Printed in the U.S.A.
Printed in the U.K. by (see last page)
ISBN: 978-3-8381-0267-2

Abstract

In the framework of this work, Scanning Capacitance Microscopy (SCM) and Spectroscopy (SCS) was applied to various silicon samples. In SCM, a tiny electrical conductive tip is scanned over a sample surface, and the capacitance between the tip an the sample is recorded. This can be used to investigate the local electrical behaviour of the sample surface. All samples used in this work showed a Metal-Oxide-Semiconductor (MOS) behaviour and the data were interpreted by using MOS theory. The measurements were performed under ambient conditions (room temperature and ambient atmosphere).

The first project was to explore the properties of ZrO_2 as dielectric material for SCM. The low growth temperature (Metal Organic Chemical Vapour Deposition, MOCVD, at 450 °C) together with the good reproducibility of the MOCVD-process and the high dielectric constant make ZrO_2 a very promising material for SCM applications. Compared with SiO_2 as dielectric material, much thicker ZrO_2 layers can be used resulting in reduced leakage currents and improved signal quality. ZrO_2 was found to be quite insensitive to parasitic charging effects, which often disturb SCM measurements on samples with (thin) SiO_2 layers.

SCM was also used to investigate Focussed Ion Beam (FIB) induced damage in silicon. The technologically important beam shape was determined by measuring the SCM image of FIB processed implantation spots and by comparison of topographical and SCM data. Further, the question was investigated how deep impinging ions generate measurable damage below the silicon surface. For this purpose, trenches were manufactured using FIB and analysed by SCM in cross sectional geometry. It turned out that SCM shows a three times higher sensitivity to ion damage than Transmission Electron Microscopy (TEM).

However, the most important part of this work was to design a setup for *quantitative* Scanning Capacitance Spectroscopy (QSCS), where an external, ultrahigh precision, true capacitance bridge is used together with a commercially available Atomic Force Microscope (AFM). A detailed description of the setup is given and the differences to conventional, *qualitative* SCM are discussed.

The setup is sensitive enough to resolve the energetic distribution of interface trapped charges. This is an advantage compared to conventional SCMs, which have a limited energy resolution due to the large applied modulation voltages. Furthermore, the setup was used to investigate a ZrO_2 layer that shows small, periodic thickness variation (ripples) on micrometer length scales at the edges of the samples due to self organization effects during the growth process. The local oxide charge density distribution of this ZrO_2 ripples was measured *quantitatively*.

Contents

1 Introduction — 5
  1.1 State of the Art 6
  1.2 Overview over the Thesis 10

2 1D MOS Theory — 13
  2.1 The MOS Capacitor 14
  2.2 Field Effect and Band Diagrams 17
  2.3 Calculation of the Total Capacitance vs. Gate Voltage . . . 23
    2.3.1 Poisson Equation at Low Frequencies 24
    2.3.2 Poisson Equation at High Frequencies 26
    2.3.3 Solving the Low Frequency Poisson Equation: The Surface Electric Field 27
    2.3.4 Doping Level and Oxide Thickness Dependence 29
    2.3.5 The Flatband Capacitance and Flatband Voltage . . . 31
  2.4 Oxide Charges 31
  2.5 Edge Effects – Deviations from the Flat Capacitor Model . . . 35

3 Equipment — 37
  3.1 The Standard Commercial Scanning Capacitance Equipment . 38
    3.1.1 Spectroscopy Mode 41
    3.1.2 Imaging Mode 43
  3.2 The Capacitance Bridge Setup 44
    3.2.1 The Capacitance Bridge 44
    3.2.2 Setup for AFM-Based Measurements with the Capacitance Bridge 47
  3.3 Establishing an Electrical Back Contact to the Sample 54
  3.4 Factors Influencing the Capacitance Signal 56
  3.5 Mechanism of Bias Dependent Contrast in Scanning Capacitance Microscopy Images 59

4 Zirconium Dioxide as Dielectric for SCM — 66
- 4.1 Introduction 66
- 4.2 Deposition of the Dielectric Layers 68
  - 4.2.1 Cleaning Procedure Prior to Oxidation or Deposition . 68
  - 4.2.2 High Temperature Oxidation Process for SiO_2 69
  - 4.2.3 Metal Organic Chemical Vapour Deposition of ZrO_2 . . 69
- 4.3 Investigating the Properties of ZrO_2 as a High Quality Dielectric for SCM 73
  - 4.3.1 Surface Roughness of ZrO_2 and Impact on the Resolution of SCM 74
  - 4.3.2 ZrO_2 Stability Against Electrical Stress 76
  - 4.3.3 Comparison of pn-Junctions Covered with SiO_2 and ZrO_2 77

5 FIB Induced Damage Investigated by SCM — 81
- 5.1 Introduction 81
- 5.2 The Focussed Ion Beam System 83
- 5.3 Sample Preparation and Experimental Considerations 85
- 5.4 Ion Beam Profile – Damage on Si-Surfaces 87
- 5.5 Ion Beam Damage Inside Si-Samples 89

6 Microscopic Capacitance Spectroscopy — 94
- 6.1 Introduction 94
- 6.2 Developing a Useful Parameter Set 94
  - 6.2.1 Pointwise Averaging – Average Time Level 94
  - 6.2.2 The Modulation Voltage 96
  - 6.2.3 Does QSCS Influence the Sample? 98
- 6.3 AFM Tip Drifts and Countermeasures 99
  - 6.3.1 Piezo Creep 100
  - 6.3.2 Thermal Expansion 101
- 6.4 Data Extraction and Analysis 108
- 6.5 C(V) Spectroscopy on Macroscopic and Nanoscopic Capacitors 111
- 6.6 Quantitative Investigation of the Properties of a ZrO_2 Layer . 118

7 Critical Remarks and Outlook — 125

A List of Symbols — 127

B List of Constants — 134

C List of Publications — 135

Chapter 1

Introduction

Ever since the first transistors were integrated into the first integrated circuit in 1960, continual development and growth in semiconductor technology has been witnessed and has led to a steady miniaturisation of semiconductor devices. This has become generally known as *Moor's Law*, which states that the amount of transistors in an integrated circuit doubles about every 18 months. Hand in hand with this exceptional miniaturisation went the demand for powerful imaging techniques that allowed to investigate the latest generation of semiconductor devices. (Far-field) optical microscopes reached the diffraction limit (about 0.5 μm) when the first devices were created with features smaller than about 0.5 μm between the years 1985 and 1990.

Electron microscopes, invented in 1931 by Ernst Ruska and Max Knoll, were used to image the small structures. Atomic resolution was reached the first time in the 1960's. However, difficult sample preparation (in Transmission Electron Microscopy, TEM), charging effects on insulating samples (Scanning Electron Microscopy, SEM) and the very expensive equipment were often annoying.

Then, in the year 1981, Gerd Binnig and Heinrich Rohrer [14] invented the Scanning Tunnelling Microscope (STM) where a metal needle is scanned over an electrically conductive sample to investigate the topography of the sample. Shortly afterwards, they showed the possibility of achieving atomic resolution with the recently invented STM. In STM, the electric current caused by the tunnelling of electrons from a conductive needle or tip to the sample is used to maintain a constant separation between tip and sample. For their achievements, Ernst Ruska, Gerd Binning and Heinrich Rohrer shared the Nobel Prize in Physics 1986. Because STM requires that the sample surface be conductive in order to enable a measurable electric (tunnel) current, atomic force microscopy (AFM) was developed in 1986 by Binning and Rohrer for measurements on surfaces that are not a good conductor [13]. To obtain nearly the same lateral resolution as in STM, AFM uses a sharp tip formed on a soft cantilever to probe the interaction (force) between the tip and the sample surface. If the force between the tip and the sample is increased, the cantilever starts to bend and a laser beam aimed at the cantilever is deflected. This deflection can be detected and can be used in a feedback loop to maintain a constant force between the tip and the sample. AFM has been developed as an independent technique and has found much more applications. Almost all kind of materials can be measured by AFM [36, 68, 91, 116, 94], not only the morphology but also many surface/interface properties (mechanical, magnetic, electric, optical, thermal, chemical properties) when using various derivations of the original AFM [64, 108, 66, 71, 11, 112]. The term *Scanning Probe Microscopy* (SPM) summarizes all methods based on the original AFM.

Out of this large number of available scanning probe methods, *Scanning Capacitance Microscopy* (SCM) and *Scanning Capacitance Spectroscopy* (SCS) are some of the most promising tools for failure analysis and quality control in semiconductor fabrication. In SCM and SCS, the tiny *capacitance* between the tip and the sample is measured to obtain capacitance maps of the sample region (in SCM) or capacitance spectra, that show the capacitance behaviour of a single sample spot under varying voltage conditions (in SCS). Conventional capacitance measurements are used routinely in semiconductor device characterization for decades and yield various important parameters such as oxide thickness, oxide charges, work functions of metals and doping levels in semiconductors. In SCM it is tried to combine this useful technique with scanning probe methods in order to obtain spatially resolved results with nm-resolution. As it turned out, however, SCM measurements are a technical challenge.

1.1 State of the Art

Scanning Capacitance Microscopy was invented in 1985 by J. Matey and J. Blanc [45] as a new mechanical method to image topographic features (rather than electrical features). Note that SCM was developed *before* the invention of the AFM in 1986. In its original design, the probe consisted of an electrode that was attached to the edge of a diamond stylus which was in mechanical contact with the sample as seen in figure 1.1. The bottom of the electrode was about 20 nm from the bottom of the stylus. A capacitance sensor was used to detect the small variations in the capacitance signal due to topographic variations. This capacitance sensor was based on the read out electronics of VideoDisc systems [88]. The probe area and therefore the

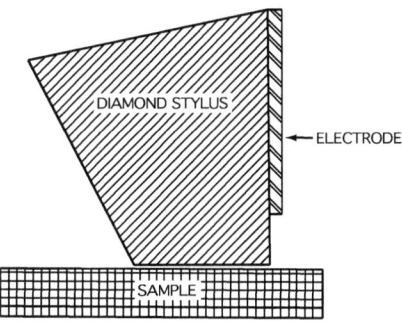

Figure 1.1: Scheme of the capacitance probe of the first SCM system. The diamond stylus is only used as a carrier for the metal electrode. Scheme was taken from J. Matey et al. [45].

approximate lateral resolution was about 0.5 μm^2.

The invention of the AFM and the development of sharp AFM tips allowed an 100-fold increase in lateral resolution. To achieve electrical conductivity of the AFM tips, either doped Si AFM probes, or Pt or PtIr covered AFM probes were used. However, the doped Si tips often suffered from depletion effects inside the tip [120], whereas the the metal coated tips suffered from abrasion and undesirable variations in SCM data. Around the year 1999, highly doped diamond coated AFM tips were introduced. Refer to figure 1.2. Due to the diamond coating, these tips are very resistant against abrasion and give a stable SCM signal. Because of the *very* high boron doping (1×10^{20} cm^{-3}), these tips do not suffer from depletion effects. The tip area of diamond coated AFM tips is below 100×100 nm^2 [2]. Currently, full diamond AFM tips are under development, which exhibit even smaller tip areas[6].

The current state of the art of SCM and SCS can be found in the review article [120, 107], and recent developments, such as SCM measurements at low frequencies, either by electric force based methods detecting higher harmonics in the AFM signal or special sensor circuits are outlined in references [61, 58].

The demand for smaller and smaller semiconductor devices is accompanied by the requirement of high resolution imaging techniques for dopant and carrier profiling. One of the main competitors in this field is SCM, beside Scanning Spreading Resistance Microscopy (SSRM) (for a comparison of methods for 2D carrier profiling refer to [26]). Several groups are investi-

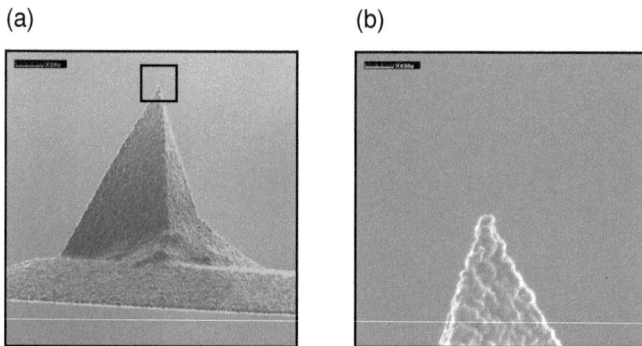

Figure 1.2: (a) Scanning Electron Microscope (SEM) image of a conductive diamond coated force modulation AFM tip (CDT-FMR) commonly used in SCM. (b) A magnification of the tip.

gating pn-junctions in order to determine the local carrier density [19, 54, 48] using SCM and Kelvin Probe Force Microscopy.

A way to increase the spatial resolution of SCM by many order of magnitudes is to use *angle-bevelling* [32, 31]. Very narrow doping profiles can be broadened geometrically by simply bevelling the sample under the correct angle to create a broadened projection of the doping profile (refer to figure 1.3)

Numerous SCM applications are already reported in the literature. Industrial devices such as MOSFETs [57] and other VLSI structures [81] and semiconductor lasers [28] are also frequently studied by cross sectional SCM. Here, the extension of the drain and source areas and the carrier concentration in the channel area are of special interest [84]. Recent publications report on SCM measurements on actively operated devices, where the local carrier concentration was studied under various operation conditions [24, 119, 25]. In this way, effective channel lengths of MOSFETs were determined [111].

Scanning Capacitance Microscopy is also used by some groups to investigate oxide charge trapping properties of SiO_2 on silicon [72, 41] with high lateral resolution. They exploit the possibility to actively *write* (inject) charges locally into the SiO_2 layer by applying higher voltages between the tip and the sample than commonly used for capacitance imaging. Afterwards the charged regions can be investigated by SCM at lower voltages. In most cases, the charging of oxide layers during SCM measurements is disadvantageous and is observed rather a parasitic effect. Unfortunately, it is not always pos-

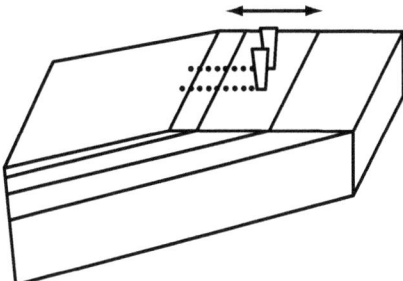

Figure 1.3: Scheme of an angle-bevelled sample. Differently doped layers appear broader than their diameter. Image was taken from the PhD-thesis of P. Eyben [30]

sible to adjust the voltages in commercial SCM equipment to values which do not lead to charging effects. Other groups try to apply SCM for routine oxide quality control at ambient temperature [21] or even at variable temperatures [56]. The variable temperature approach makes it possible to obtain a more detailed insight into the physical processes that govern the trapping and de-trapping of electrons or holes inside the oxide or at the interface between the oxide and silicon.

Besides silicon, other semiconductor systems were also investigated by SCM. E.g. a Metal-Insulator-Semiconductor (MIS) behaviour was found by Richter et al. [95] on InP. InAs self assembled quantum dots on GaAs were investigated by Yamamoto et al. [42] using the same capacitance bridge introduced in this work for studies of silicon samples. Figure 1.4 shows the topographic and capacitance images of an InAs quantum dots recorded by Yamamoto et al. On GaInP, ordered and disordered domains [50] were imaged.

Although applied successfully for various applications, SCM is not an easy and strait forward technique to use. In detail, quantitative and reproducible measurements are a serious problem, since sample preparation has a dramatic influence on the results especially in cross sectional measurements. In the literature, many articles can be found on the subject of sample preparation [33, 51, 100, 110, 119]. In addition to the technical problems with sample preparation, the physical processes leading to contrast in SCM images are also not fully understood. It was shown for unipolar doped samples that the contrast is monotonic only if the tip voltage is adjusted properly (refer to chapter 3.5 or the articles [98, 105]). In addition to that, other groups

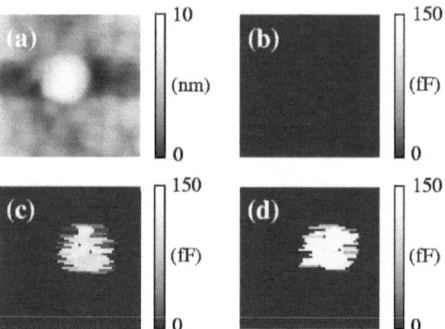

Figure 1.4: (a) Topographic AFM image of an InAs quantum dot. (b) – (d) Capacitance images of the quantum dot recorded at different tip voltages (0 V, 0.5 V and 1 V, respectively) with an *Andeen Hagerling* capacitance bridge. Images were taken from Yamamoto et al. [42].

have shown that too many surface states can lead to contrast reversal and that a good oxide quality is essential for this reason [27]. Quantitative SCM measurements with high spatial resolution are still a challenge for many technical and physical reasons. First, the capacitance between the AFM tip and the sample is in the aF regime only. To obtain a reasonable signal size at reasonable data collection speed for imaging, lock-in techniques are normally used. Thus, commercial SCM systems usually yield qualitative data only. To obtain quantitative results, these data have to be calibrated. This, however, is technically complicated due to large difficulties with the reproducibility of the reference sample preparation process. In addition, the currently commercially available SCM systems do not operate at small signal conditions, which further complicates the data evaluation and makes sophisticated simulation methods almost inevitable [117, 47, 60, 49], especially if structures in the nm-regime have to be investigated.

1.2 Overview over the Thesis

In this work, *Scanning Capacitance Microscopy* and *Scanning Capacitance Spectroscopy* were applied on silicon samples covered with SiO_2 and ZrO_2. An advanced experimental setup (with the AH2550A capacitance bridge) was developed that allows for *quantitative* capacitance measurements, and which complements the qualitative investigations with the conventional SCM module.

The insights and operational experience gained during the experiments are summarized. This work was also created as a reference for future projects based on SCM. Some topics already discussed in the thesis, were again summarized elsewhere in the thesis to increase the readability.

Chapter 2 develops the basic theoretical knowledge needed to understand the measurement principle of SCM. It gives an introduction into 1D MOS theory and explains the deviations from this oversimplified model due to the geometry of the SPM tip.

In **chapter 3**, the equipment for the following investigations is described. Efforts are made to clearly distinguish between the two experimental setups for capacitance spectroscopy:

- **Quantitative Scanning Capacitance Spectroscopy (QSCS) with the AH2550A true capacitance bridge at 1kHz**

- **(Non-quantitative) SCM and SCS with the commercial, conventional SCM module at about 1GHz**

A summary of the basic measurement principles of both the AH2550A capacitance bridge as well as the SCM module is given. The various effects that influence the outcome of capacitance measurements with both setups are described in chapter 3.4.

Chapter 4 deals with the substitution of the commonly used SiO_2 layer with a high-ε ZrO_2 layer (in literature often refered to as *high-k* layers) as a dielectric for conventional SCM investigations. The differences in the topographic structure and in the electrical behaviour are discussed. It turns out that ZrO_2 layers are superior to SiO_2 layers for many SCM applications.

In **chapter 5**, conventional SCM is used as a very sensitive method to determine the extent of damage accumulated during the local irradiation of a Si sample with Ga^+-ions inside a Focussed Ion Beam (FIB) system. As it turned out, conventional SCM is three times more sensitive at imaging ion-damage than Transmission Electron Microscopy (TEM).

In **chapter 6**, the measurements performed with the AH2550A capacitance bridge setup are presented. In the first part, useful measurement parameters are defined and some parasitic effects are described that are inherent in the measurement setup. In the second part, the first measurements on SiO_2 and ZrO_2 covered Si samples are presented. It turned out that the spectroscopic resolution of the AH2550A capacitance bridge setup is superior to the resolution of the conventional SCM module. After "the prove of principle", the setup was applied to investigate the *quantitative* distribution of oxide charges on a ZrO_2 covered sample.

Finally, **chapter 7** gives some critical remarks on the presented work and an outlook to possible future projects and applications.

Chapter 2

1D MOS Theory

Due to its many applications, the Metal-Oxide-Semiconductor (MOS) system is of tremendous importance in todays microelectronics. MOS systems are applied mainly as MOS Field Effect Transistors (MOSFETs), flash memory devices and in Charge Coupled Devices (CCD). A large amount of the semiconductor industry is involved in the production of the different MOS based devices.

The principle of the field effect was discovered in 1935 by Lilienfeld[65] and Heil[40]. An enormous amount of research has been conducted on MOS devices since the the first proposal of a MOS capacitor as a voltage dependent capacitor in 1959 [79, 89] and the first implementation of a MOSFET by Kahng and Atalla [53] in 1960.

As Scanning Capacitance Microscopy on silicon samples lead to the creation of a MOS capacitor consisting of the AFM tip, an oxide layer and the silicon sample, a detailed comprehension of the physics of MOS systems is required for the work presented here. This chapter is meant to impart this knowledge.

The theory which is presented here and which is applied later in this work is only valid for large, flat capacitors where so called *edge effects* due to fringing electric fields at the capacitor edge can be neglected. Note that an AFM tip is *not* a large capacitor anymore and edge effects are an issue. The relevant deviations of the flat capacitor model that are introduced by the small AFM tip are discussed in chapter 2.5.

The symbols used in the following equations are most often equivalent to Nicollian and Brews [83]. Whereas in theoretical calculations it is common practice to use capacitance and charges *per unit area* to obtain values independent of the capacitor- or tip area (which is consistent with Nicollian and Brews [83]), in experimental analysis it is more advantageous to use the non-normalised values. To avoid confusions and for an easy distinction, the non-normalised values are *primed*. Therefore, e.g. the values $\overline{C}_s$, $\overline{C}_{ox}$, $\overline{Q}_s$, $\overline{Q}_{it}$, ..., mean capacitance or charge in the common sense, whereas the values C_s, C_{ox}, Q_s, Q_{it}, ..., mean capacitance or charge *per unit area*. For comparison with literature, the oxide charge densities can be calculated in real *charge densities* or *number densities*. Therefore, the interface trapped charge density Q_{it} is given in units of C/cm^2 (Coulomb per area), whereas the *number* of charges per area N_{it} is calculated by division with the the electron charge q: $N_{it} = Q_{it}/q$ (in cm^{-2}). Unfortunately, in scientific literature the various terms Q_x for oxide charge densities are used equivalently for charge densities (Coulomb/cm^2) as well as for number densities (1/cm^2).

In the discussion of the energy band diagrams, the terms *energy level*, *potential* and *voltage* are used synonymously. The exact definition of the energy levels and when the "conversion factor" q has to be applied can be taken from the illustrations of the band diagrams. For example, the energy difference of the work functions of the gate and the semiconductor is $q \cdot W_{ms}$ (in eV or Joule) whereas the flatband voltage depends on the work function difference W_{ms} measured in Volt. In addition, the terms *gate voltage* and *tip voltage* have the same meaning. However, *gate voltage* is used throughout the theory part, whereas in the experimental parts the term *tip voltage* seems to be more convenient.

After an introduction into the subject in chapter 2.1, some basic knowledge is delivered in chapter 2.2. Chapter 2.3 schematically describes how the capacitance of a MOS capacitor can be calculated in the high frequency as well as in the low frequency case, and derives some other useful expressions important for this work. An introduction into the different types of oxide charges and how they influence the capacitance behaviour of a MOS system is given in chapter 2.4. Finally, the deviations due to edge effects and the limit of the flat capacitor model is discussed in chapter 2.5. Most of the deviations presented in these chapters is based on [83].

2.1 The MOS Capacitor

Traditionally, a MOS capacitor consists of an oxide layer usually thermally grown on a silicon substrate and a *gate* electrode made by vacuum deposition of a metal or by the deposition of polysilicon (figure 2.1 (a)). In the case of this work, the gate electrode of the MOS capacitor consists of the electrical conductive, diamond coated AFM tip (figure 2.1 (b)). The back-contact is also an issue, because under certain circumstances it has an impact on

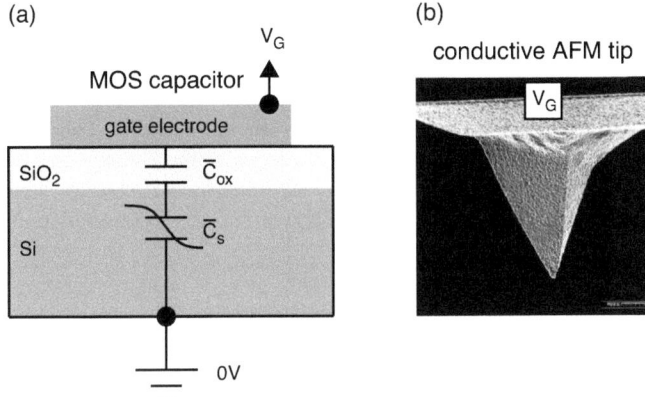

Figure 2.1: (a) Scheme of a MOS capacitor. (b) In case of this work, the gate electrode is replaced by a conductive AFM tip.

the behaviour of the MOS system. This will be discussed in chapter 3.3. In the work presented here, the oxide layer consists solely of SiO_2 or ZrO_2. However, the subjects discussed in chapter 2 are valid also for other types of dielectrics.

Unlike a simple capacitor consisting of two metal electrodes separated by a dielectric layer, the total capacitance $\overline{C}$ of a MOS capacitor varies with the applied voltage. The total capacitance $\overline{C}$ consists of the capacitance $\overline{C}_{ox}$ of the oxide layer which is given by:

$$\overline{C}_{ox} = \varepsilon_{ox} \cdot \varepsilon_0 \cdot \frac{A}{d} \quad \text{(unit: F)} \tag{2.1}$$

Here, A is the area of the capacitor (or the effective area of the AFM tip), d is the thickness of the oxide layer, and ε_{ox} is the dielectric constant of the oxide.

$\overline{C}$ depends also of the capacitance $\overline{C}_s$ of the silicon layer immediately below the SiO_2 layer, which eventually is depleted from charge carriers. The capacitance $\overline{C}_s$ will be called *silicon* or *semiconductor capacitance* from now. The depleted area inside the silicon will vary with the voltage applied to the MOS system. In the following discussion of the behaviour of the MOS system, this voltage will be called gate voltage V_G, although most often in this work, the top electrode of the MOS system consists of a conductive AFM tip. The silicon capacitance depends on the gate voltage V_G, because at certain values of V_G, the gate will repel the majority charge carriers of the bulk silicon near the SiO$_2$ layer: $\overline{C}_s = \overline{C}_s(V_G)$. The silicon region below the SiO$_2$ layer is *depleted* from majority carriers, and behaves like an insulator. This is the case for example on a p-doped samples where positive voltages are applied to the gate with respect to the doped sample. The *depletion layer width w* is the extent of the depletion layer inside the silicon. Similar to equation 2.1, the silicon capacitance $\overline{C}_s$ (in units Farad) can be calculated if the depletion layer width w, the dielectric constant of the semiconductor ε_s and the area A is known.

$$\overline{C}_s = \varepsilon_s \cdot \varepsilon_0 \cdot \frac{A}{w} \qquad \text{(unit: F)} \tag{2.2}$$

It is now possible to calculate the total capacitance $\overline{C}$ of the MOS system. Figure 2.1 indicates that the total capacitance can be calculated by the oxide capacitance $\overline{C}_{ox}$ in series with the silicon capacitance $\overline{C}_s$:

$$\frac{1}{\overline{C}} = \frac{1}{\overline{C}_{ox}} + \frac{1}{\overline{C}_s} \tag{2.3}$$

For thin oxides and low doped samples, the total capacitance will be dominated by the silicon capacitance $\overline{C}_s$ (refer to chapter 2.3.4, figure 2.7 (a)).

It is important for the following discussion to understand that two conceptually different capacitance expression can be defined. The *static* capacitance is defined as:

$$\overline{C}_{stat} = \frac{\overline{Q}}{V_G} \tag{2.4}$$

Here, $\overline{Q}$ is the total charge on the capacitor in Coulomb and V_G is the applied voltage. The unit of $\overline{C}_{stat}$ therefore is Farad. The *differential* capacitance is defined as:

$$\overline{C}(V_G) = \frac{d\overline{Q}(V)}{dV}\bigg|_{V_G} \tag{2.5}$$

Again, $\overline{Q}$ is given in Coulomb and $\overline{C}(V_G)$ is given in Farad. Both definitions can be applied to the MOS capacitor, however, the two capacitances will be different because charge on an MOS capacitor can vary *non-linearly* with the voltage. Therefore, of the two capacitances, the most important in MOS capacitor *measurements* is the differential capacitance defined in equation 2.5. To measure the total capacitance as a function of the applied gate voltage, a small AC *excitation voltage* V_{excit} is superimposed on the DC gate voltage V_G, as is shown in figure 2.2. Measuring the capacitance $\overline{C}(V_G)$ at different V_G values gives a so called C(V) curve of the MOS capacitor.

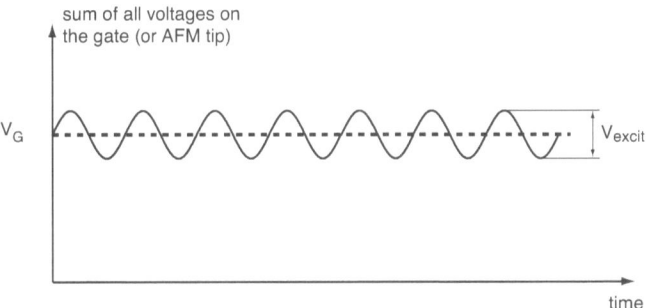

Figure 2.2: Principle of measuring the differential capacitance $\overline{C}(V_G)$ by superimposing a small variable (AC) excitation voltage V_{excit} to the constant voltage bias V_G.

The measured C(V) curve is a good approximation to the ideal differential capacitance in equation 2.5 only if the excitation voltage is small, or, in other words, the excitation voltage has to be within the *small signal regime*. If the excitation voltage is too large, the measured curves are broadened because of the lack of a linear response to the applied excitation voltage.

The following calculations and deductions only refer to the differential capacitance for an easy comparison with experiments. Throughout this work, the term capacitance is used to mean differential capacitance.

2.2 Field Effect and Band Diagrams

To develop mathematical expressions for the behaviour of the MOS capacitor under applied voltage, first one has to understand what happens if a gate material (or AFM tip) is connected to an oxide covered Si sample. The processes that take place are best described by drawing the energy band diagrams of the involved material layers. Figure 2.3 (a) shows such a band diagram for a MOS capacitor directly after "imaginary" connecting a layer of a gate material to the oxide and avoiding charge-transfer by electrical contact between gate and the silicon bulk material. The oxide layer is assumed to be a perfect insulator, there are no leakage currents through the oxide. In general, the work function of the gate electrode qW_m will be different from the work function of the silicon sample qW_s. The work function of a material is the energy needed to excite an electron from the Fermi level (E_{Fm} and E_{Fs} for the gate material and the silicon sample, respectively) to the vacuum level, where the

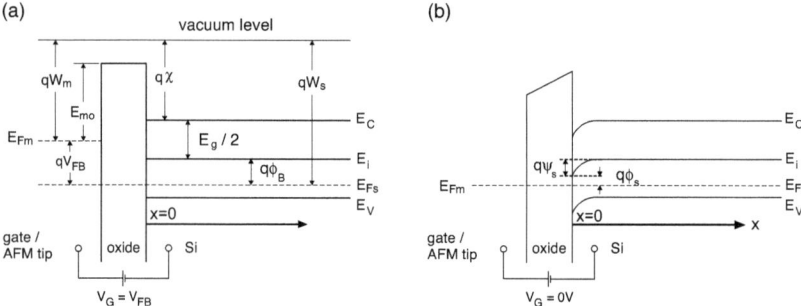

Figure 2.3: The band diagram of a MOS capacitor (a) under flatband conditions and (b) at zero gate voltage.

electron is not bound to the according material anymore. The case shown in figure 2.3 (a) is referred to as the *flatband condition*, because the conduction as well as the valence bands are flat, that is the band edge energies E_V and E_C are constant all over the silicon. However, this state is not the thermodynamic equilibrium of the MOS system (the Fermi level is a thermodynamic potential which has to be constant throughout the observed system in thermodynamic equilibrium). In reality, very small but finite leakage currents will equilibrate the Fermi levels. Therefore, the flatband condition can only be maintained by applying an external voltage to counter the tendency of the charge carriers to find an equilibrium distribution. This voltage is referred to as the *flatband voltage* V_{FB}. In case of a short circuit (or, more realistic, due to non-zero leakage currents) between the gate and the semiconductor, the electrons will flow from the gate to the silicon. However, this undermines charge neutrality on both sides of the dielectric layer and an electric field builds up between the gate and the sample that prevents further net flow of electrons. The equilibrium is reached if the drift rate of charge carriers due to the electric field counters the tendency of the carriers to occupy the lowest energy levels. In equilibrium, the Fermi level is the same everywhere, which is shown in figure 2.3 (b).

The electric field and the correlated potential drop is distributed between the dielectric layer (the SiO_2 or ZrO_2 layer in the case of this work) and a silicon layer of varying extent. Inside the silicon, this leads to a bending of the valence and conduction bands edges (E_V and E_C respectively) near the Si-SiO$_2$ interface. Figure 2.4 shows the electric field of a MOS system. The gate electrode or AFM tip is assumed to be a perfect conductor, therefore

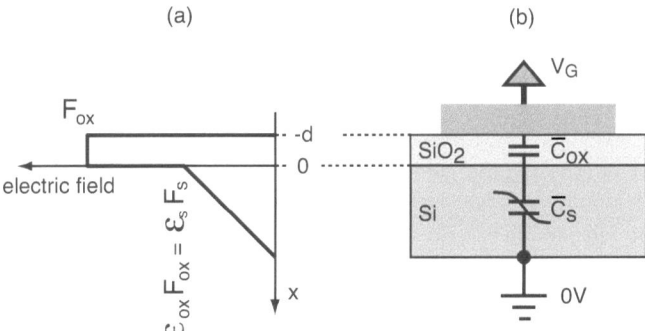

Figure 2.4: The electric field inside the MOS capacitor. The discontinuity at the interface is due to the different dielectric constants of the oxide and the silicon sample.

the field does not penetrate the gate or AFM tip.

Some energy level inside the silicon depend on the distance x from the oxide. Therefore, some important relations between the different energy levels and potentials in figure 2.3 will be defined now: The relation between the Fermi level E_{Fs} and the intrinsic Fermi level $E_i(x)$ in the silicon is given by

$$q \cdot \phi(x) = E_{Fs} - E_i(x) \qquad (2.6)$$
$$\phi_s = \phi(0) \quad \text{at the Si–SiO}_2 \text{ interface} \qquad (2.7)$$
$$\phi_B = \phi(\infty) \quad \text{deep inside bulk Si} \qquad (2.8)$$

The intrinsic Fermi level is the Fermi level in case of an undoped semiconductor. At room temperature, it shows a negligible temperature dependence and lies approximately at midgap, at the same energy distance to the conduction and the valence band edge. The *surface potential* ϕ_s occurs at the surface of the semiconductor ($x = 0$) and the *bulk potential* ϕ_B occurs inside the semiconductor bulk.

The band bending $\psi(x)$ quantitatively describes the bending of the valence and conduction bands at any point x in the silicon with respect to the bulk value:

$$\psi(x) = \phi(x) - \phi_B \qquad (2.9)$$

CHAPTER 2. 1D MOS THEORY

$$\psi_s = \psi(0) \quad \text{at the Si-SiO}_2 \text{ interface} \quad (2.10)$$

The band bending at the silicon surface ψ_s is commonly referred to as the *barrier height* [83], page 51. Note that ψ_s is the total potential difference between the silicon surface and the bulk silicon.

The Fermi level E_{Fs}, and subsequently the work function qW_s of the silicon depend on the doping concentration of N_A or N_D (in units cm^{-3}) of acceptors or donators, respectively. The work function can be calculated by summing up the corresponding potentials in figure 2.3:

$$W_s = \chi + \frac{E_g}{2 \cdot q} - \phi_B \quad (2.11)$$

χ is the electron affinity and E_g the band gap of silicon. Experimental values for the electron affinity and the band gap can be taken from literature [109] ($q \cdot \chi = 4.05$ eV, $E_g = 1.12$ eV at 300K). The work function difference W_{ms} between the gate material and the silicon is very important for the behaviour of the MOS system and is defined by:

$$W_{ms} = W_m - W_s \quad (2.12)$$

The bulk potential ϕ_B now gives the dependency on the doping level N_A or N_D:

$$\phi_B = \frac{k \cdot T}{q} \cdot \ln\frac{N_D}{n_i} \quad \text{n-type} \quad (2.13)$$

$$\phi_B = \frac{k \cdot T}{q} \cdot \ln\frac{n_i}{N_A} \quad \text{p-type} \quad (2.14)$$

n_i is the intrinsic carrier concentration in case the semiconductor is undoped. It corresponds to the intrinsic Fermi level E_i. An experimental value for n_i was again taken from literature [109] : $n_i = 1.45 \times 10^{10}$ cm^{-3} at 300K [1].

Due to equations 2.6 and 2.9, also the surface potential ϕ_s and the band bending ψ_s depend on the silicon doping level via the Fermi level of silicon E_{Fs}.

From figure 2.3 (a) and (b) one can see that the energy levels and the band bending depend on the applied voltage. This is called the *field effect*. Lets

---
[1] Due to the logarithm in equation 2.13 and 2.14, the bulk potential ϕ_B and subsequently the work function qW_s is insensitive to measurement errors of the intrinsic carrier density n_i. Even varying n_i one decade in both directions does alter the work function only by less than 0.05 eV.

explore what happens if gate voltages other than $V_G=0$ or $V_G=V_{FB}$ are applied on the MOS capacitor. Based on the amount of electrons and holes in the silicon layer below the oxide, there are three possible cases, which will be discussed for a p-type silicon semiconductor:

- **Accumulation:** If one applies a gate voltage that is *lower* than the flatband voltage, $V_G \ll V_{FB}$, the energy bands behave like in figure 2.5 (a). No electrical current flow is possible through the oxide layer, therefore the Fermi level is constant throughout the silicon. The very large negative voltage on the gate contact makes the holes of the silicon sample to accumulate near the Si-SiO$_2$ interface. The majority carrier concentration (holes) is increased significantly. In the band diagram of figure 2.5 (a) this means that the band bending ψ_s increases and that the conduction and valence bands are bent upwards. Because a lot of charge carriers are available near the Si-SiO$_2$ interface, the total capacitance of the system is the capacitance of the oxide: $\overline{C} = \overline{C}_{ox}$.

- **Depletion:** If the gate voltage is larger than the flatband voltage, $V_G > V_{FB}$, the majority carriers (holes) will be repelled from the silicon surface. The result is a region depleted from charge carriers. This *depletion zone* acts like a dielectric layer of thickness w with a dielectric constant ε_s of silicon. The capacitance $\overline{C}_s$ of the depletion layer can then be calculated by using equation 2.2. The larger the depletion layer width w, the lower is the capacitance. The resulting capacitance of the system can be calculated by summing up the reciprocal values of the oxide capacitance $\overline{C}_{ox}$ and the depletion layer capacitance $\overline{C}_s$ as shown in equation 2.3. In depletion, the total capacitance of the MOS capacitor will be lower than the oxide capacitance.

- **Inversion:** Finally, if the gate voltage is further increased ($V_G \gg V_{FB}$), the energy bands are further bent downwards and the intrinsic level E_i falls below the Fermi level. Now, there are more minority carriers (electrons) than majority carriers near the interface. This is called an *inversion layer*. The inversion layer is followed by a depletion layer, because the band bending ψ_s steadily goes to zero with increasing distance x from the interface. The impact of the inversion layer on the total capacitance $\overline{C}$ depends on the frequency of the applied excitation voltage. At low frequencies the inversion layer acts as the bottom electrode, and, like in accumulation, the total capacitance is governed only by the oxide capacitance $\overline{C} = \overline{C}_{ox}$. This case is described in chapter 2.3.1. At high frequencies, the concentration of the minority carriers inside the inversion layer does not follow the quick modulation

CHAPTER 2. 1D MOS THEORY

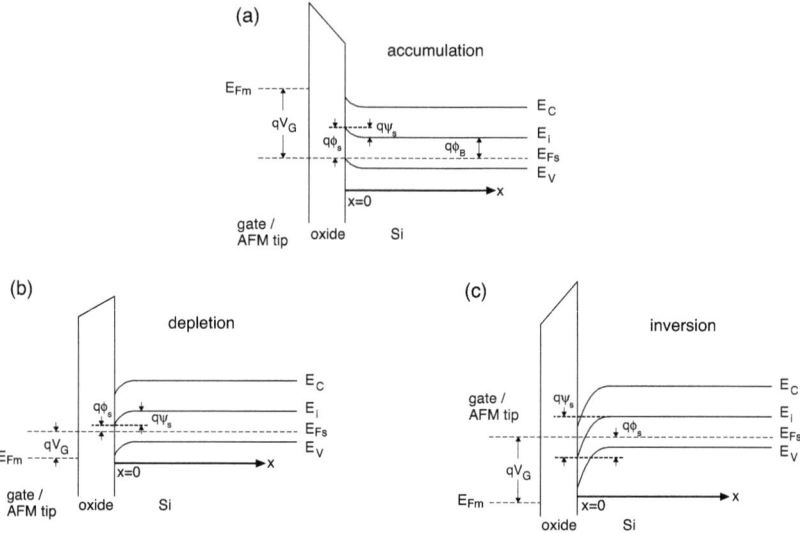

Figure 2.5: Band diagrams of a MOS system with p-type semiconductor. (a) In accumulation, the bands bend up and majority carriers (in this case: holes) accumulate at the silicon - SiO$_2$ interface. (b) In depletion, the silicon layer near the interface is depleted from free charge carriers. (c) In inversion, the minority carrier density (electrons) at the interface is larger than the majority carrier density (holes).

of the gate voltage and the total capacitance is governed also by the depletion layer like in depletion. The high frequency case is discussed in chapter 2.3.2.

The width of the depletion layer depends on the applied band bending and is given for a p-type semiconductor:

$$w = \sqrt{\frac{2 \cdot \psi_s \cdot \varepsilon_0 \cdot \varepsilon_s}{q \cdot N_A}} \quad (2.15)$$

The higher the doping level (acceptors N_A) of a semiconductor the smaller is the depletion layer width w. ε_s is the dielectric constant of the semiconductor. This is also important when considering the lateral resolution of SCM. The same is true for n-type semiconductors, if acceptor concentration N_A is substituted by donor concentration N_D.

Two other values are very important in the following discussion: the *intrinsic Debye length* λ_i and the doping type dependent *extrinsic Debye lengths* λ_p and λ_n. These are characteristic lengths over which the carrier density in a semiconductor changes by a factor e (Euler's Number), and are given by

$$\lambda_i = \sqrt{\frac{\varepsilon_0 \cdot \varepsilon_s \cdot k \cdot T}{q^2 \cdot n_i}} \tag{2.16}$$

$$\lambda_p = \sqrt{\frac{\varepsilon_0 \cdot \varepsilon_s \cdot k \cdot T}{q^2 \cdot N_A}} \quad \text{p-type} \tag{2.17}$$

$$\lambda_n = \sqrt{\frac{\varepsilon_0 \cdot \varepsilon_s \cdot k \cdot T}{q^2 \cdot N_D}} \quad \text{n-type} \tag{2.18}$$

At the depletion layer edge, the transition from depletion to the carrier density in the bulk is occurring over a distance comparable to the extrinsic Debye length.

2.3 Calculation of the Total Capacitance vs. Gate Voltage

To calculate the total capacitance versus gate voltage [C(V)] curves in the low frequency as well as in the high frequency case, one has to start with the Poisson equation in one dimension [109]:

$$\frac{\mathrm{d}^2 \psi(x)}{\mathrm{d}x^2} = \frac{\rho(x)}{\varepsilon_0 \cdot \varepsilon_s} \tag{2.19}$$

For the following deviations, it does not matter whether the band bending ψ_s or the surface potential ϕ_s is used, because both values only differ by an additive constant (ϕ_B, refer to equation 2.9). Both values describe the same physics, and define the behaviour of the MOS system by affecting the carriers in the semiconductor. The correlation to the gate voltage is very simple and will be given later. The charge density $\rho(x)$ (Coulomb/cm^3) at a certain distance x from the interface is the sum of all the fixed and free charges densities inside the semiconductor :

$$\rho(x) = q \cdot [p(x) - n(x) + N_D - N_A] \tag{2.20}$$

here $n(x)$ and $p(x)$ are the densities of electrons and holes at a location x, and N_D and N_A are the densities of ionized donator and acceptor atoms [2].

---

[2] Only uniformly doped semiconductor samples are considered in the deviations presented here.

CHAPTER 2. 1D MOS THEORY   24

Lets suppose again a p-type semiconductor, which means that the majority carriers are holes, and the minorities are electrons. The difference between low frequency and high frequency behaviour occurs in inversion. At the low frequency case, the total *number* of electrons (minorities) changes with the sum of the DC component of the gate voltage *and* the excitation voltage. At the high frequency case the total number of electrons are only defined by the DC component of the gate voltage. However, the fixed number of electrons in the inversion layer react to the high frequency excitation voltage by periodically changing the extent of the inversion layer. A more detailed discussion will be given in chapter 2.3.1 and 2.3.2

The next step is to integrate the Poisson equation once with respect to x, to obtain an expression for the electric field F_s (in common units V/m) at the silicon surface. The electric field can be converted to the total charge *per unit area* Q_s at the silicon surface by using Gauss' Law

$$Q_s = \varepsilon_0 \cdot \varepsilon_s \cdot F_s \tag{2.21}$$

where ε_s is the dielectric constant of the semiconductor. Based on equation 2.5, the differential capacitance of the silicon can be calculated at a voltage ψ_s:

$$C_s \equiv C_s(\psi_s) = \left.\frac{dQ_s(V)}{dV}\right|_{\psi_s} \tag{2.22}$$

C_s is given in Farads per unit area, because of the unit of the silicon charge density Q_s. To calculate the silicon capacitance at a given gate voltage V_G the voltage drop inside the oxide layer has to be taken into account:

$$V_G = -\frac{Q_s}{C_{ox}} - \psi_s \tag{2.23}$$

With this equation it is possible to substitute the band banding ψ_s in equation 2.22 to obtain $C_s(V_G)$. The total capacitance as function of the DC gate voltage, $C(V_G)$, can then be calculated with equation 2.3.

A comparison of the results in the low frequency and the high frequency case can be seen in figure 2.6. The low and high frequency curve only differ in the inversion regime, where the minority charge carriers dominate. In accumulation and depletion the behaviour is the same.

2.3.1 Poisson Equation at Low Frequencies

In the low frequency case, the change of the applied gate voltage due to the excitation voltage occurs so slowly, that the *number* of the minority carriers can change periodically with the excitation voltage. Furthermore,

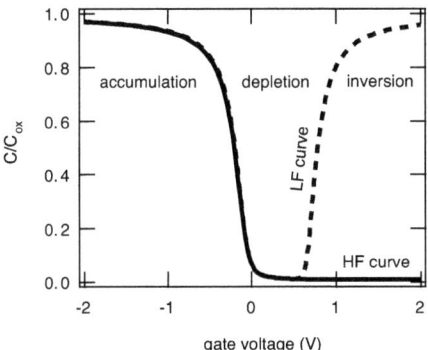

Figure 2.6: C(V) curves of an ideal MOS system with a p-type semiconductor. The low and high frequency curves only differ in the inversion region, where the minority charge carriers dominate. The curves where calculated for an acceptor concentration of $N_A = 1 \times 10^{15}$ cm^{-3}. The dielectric parameters were $d=3$ nm and $\varepsilon_{ox}=3.9$ (SiO$_2$). All curves were calculated assuming a vanishing work function difference between the gate electrode and the semiconductor ($V_{FB}=0$ V) and no interface trapped charges.

all charge carriers have enough time to redistribute. On every location inside the semiconductor, the densities of minority as well as of majority carriers can be described by a single parameter, the Fermi level. Prior to finding expressions for the charge carrier densities, some abbreviatory expression are given now:

$$u(x) = \frac{q \cdot \phi(x)}{k \cdot T} \quad (2.24)$$

$$v(x) = \frac{q \cdot \psi(x)}{k \cdot T} \quad (2.25)$$

For convenience, both the expression $v(x)$ and $u(x)$ will be called band bending from now, because they describe the same physics. Similarly, u_B and v_B mean the value in the bulk ($x = \infty$), and u_s and v_s are the values at the semiconductor surface ($x = 0$). Using the abbreviation of equation 2.24 and 2.25, the charge densities of equation 2.20 are given by:

$$n(x) = n_i \cdot \exp[u(x)] = N_D \cdot \exp[v(x)] \quad (2.26)$$
$$p(x) = n_i \cdot \exp[-u(x)] = N_A \cdot \exp[-v(x)] \quad (2.27)$$

Note that the electron and hole densities $n(x)$ and $p(x)$ strongly depend on the band bending $v(x)$ or $u(x)$ (Maxwell-Boltzmann relations). Therefore, subsequently these values depend on the applied gate voltage. The acceptor and donator concentrations N_A and N_D can be expressed in terms of u_B by using equations 2.13, 2.14 and 2.24. Insertion of the expression for the charge densities in the Poisson equation 2.19 and applying the abbreviations of equation 2.24 leads to a dimension less form of the Poisson equation:

$$\frac{d^2 u(x)}{dx^2} = \lambda_i^{-2} \cdot [\sinh u(x) - \sinh u_B] \qquad (2.28)$$

The complicated nature of this equation is due to the mutual interdependence of carrier density on potential through equations 2.26 and 2.27 and the dependence of the potential on free carrier density through Poisson's equation. However, in contrast to the poisson equation of the high frequency case (refer to chapter 2.3.2), the low frequency equation 2.28 can be integrated easily. This is performed in chapter 2.3.3.

2.3.2 Poisson Equation at High Frequencies

At high frequencies, the minority carriers only react to the excitation voltage by redistribution. The period of the excitation voltage is too short that the overall number of minorities inside the inversion layer can change significantly. In other words, the areal density of the minorities is fixed, whereas the volume density can change due to the changing extent of the inversion layer during a period of the excitation voltage. Minority and majority charges are not in equilibrium, therefore the densities of the carriers can not be described anymore by a single Fermi level. Instead, a quasi-Fermi level for the minorities is introduced. The quasi-Fermi level is time dependent, according to the AC excitation voltage. Again, the deviations will be done for a p-type semiconductor, therefore the minorities are electrons, and the (time dependent) quasi-Fermi level is termed E_{Fn}.

Lets start again by determining the charge densities per unit volume of the electrons, holes and the impurities. The expression for the electron density (minorities) in the high frequency case is different from the low frequency case (equation 2.26) and includes contributions from the quasi-Fermi level:

$$n(x) = n_i \cdot \exp[v(x) + u_{Fn}] \qquad (2.29)$$

where $v(x)$ is defined similar to equation 2.25 as $v(x) = (q\psi(x)/kT)$, and the quasi-Fermi level is included in

$$u_{Fn} = \frac{E_{Fn} - E_i}{k \cdot T} \qquad (2.30)$$

CHAPTER 2. 1D MOS THEORY

In the following, u_{Fn} is assumed to be spatially uniform throughout the semiconductor, which introduces negligible error ([83] page 158, [8]). The hole density is still given by 2.27. Finally, the Poisson equation is:

$$\frac{d^2v(x)}{dx^2} = \lambda_n^{-2} \cdot [1 - \exp(-v(x)) + \exp(v(x) + u_{Fn} + u_B)] \quad (2.31)$$

It also takes into account the redistribution of electrons at the silicon surface in response to the excitation voltage.

Basically, the high frequency C(V) curve is obtained from equation 2.31 by following the procedure of solving equations 2.21, 2.22 and 2.23 with the assumption that the total inversion layer charge (or inversion layer charge per unit *area*) is fixed by the DC gate voltage and does not change in response to the excitation voltage. The tedious and difficult calculations can be found in [83], pp. 160.

2.3.3 Solving the Low Frequency Poisson Equation: The Surface Electric Field

The calculation of the low frequency C(V) curve from the corresponding Poisson equation 2.28 is quite straightforward. One only has to integrate the Poisson equation to obtain an expression for the electric field F_s. The appropriate boundary conditions for the integration are $u(x) = u_s$ at the semiconductor surface ($x = 0$) and $u(x) = u_B$ in the bulk semiconductor ($x = \infty$). The last relation ensures charge neutrality in the bulk. To integrate the Poisson equation 2.28 the following integrating factor is introduced:

$$\frac{d}{dx}\left(\frac{du(x)}{dx}\right)^2 = 2 \cdot \frac{du(x)}{dx}\left(\frac{d^2u(x)}{dx^2}\right) \quad (2.32)$$

With the integration factor 2.32, the Poisson equation can be transformed into term that can be integrated easily. Multiplying the Poisson equation 2.28 with $2 \cdot (du(x)/dx)$, one obtains:

$$2 \cdot \frac{du(x)}{dx}\left(\frac{d^2u(x)}{dx^2}\right) = 2 \cdot \frac{du(x)}{dx} \lambda_i^{-2} [\sinh u(x) - \sinh u_B] \quad (2.33)$$

Using the integrating factor 2.32 in equation 2.33, and applying integration signs to the result leads to:

$$\int_{x=0}^{x=\infty} \frac{d}{dx}\left(\frac{du(x)}{dx}\right)^2 dx = \int_{x=0}^{x=\infty} 2 \cdot \frac{du(x)}{dx} \lambda_i^{-2} [\sinh u(x) - \sinh u_B] \, dx \quad (2.34)$$

The integration is performed from the surface ($x = 0$) to the bulk ($x = \infty$) on both sides of the equation. Performing different coordinate transformation on both sides of equation 2.34 by simply cancelling the dx terms on both sides of the equation individually, leads to:

$$\int_{(du(x)/dx)^2=(du_s/dx)^2}^{(du(x)/dx)^2=0} d\left(\frac{du(x)}{dx}\right)^2 = \int_{u(x)=u_s}^{u(x)=u_B} 2 \cdot \lambda_i^{-2} \left[\sinh u(x) - \sinh u_B\right] du(x) \quad (2.35)$$

The integration boundaries have changed from lengths values x in equation 2.34 to $(du(x)/dx)^2$ values on the left side and band bending values $u(x)$ on the right side of equation 2.35. The integration from surface value u_s to the bulk value u_B in the band bending space on the right side of equation 2.35 is consistent with the integration from the surface $x = 0$ to the bulk $x = \infty$ in the length space of equation 2.34. The boundary conditions of the integral on the left side of equation 2.35 are proportional to squared electric field values F^2 (in $(V/m)^2$) due to the relation:

$$\left(\frac{du(x)}{dx}\right)^2 = \left(\frac{q \cdot F}{k \cdot T}\right)^2 \quad (2.36)$$

Remember, the band bending $u(x)$ is nothing else than a (dimensionless) potential (refer to equation 2.24), and differentiation with respect to x will result in a value proportional to an electric field. At the semiconductor surface, the electric field will be F_s, whereas in the bulk the field is zero due to charge neutrality. Using equation 2.36 for another coordinate transformation, one can rewrite the left side of 2.35 again:

$$\left(\frac{q}{k \cdot T}\right)^2 \int_{F^2=(F_s)^2}^{F^2=0} d(F^2) = \int_{u(x)=u_s}^{u(x)=u_B} 2 \cdot \lambda_i^{-2} \left[\sinh u(x) - \sinh u_B\right] du(x) \quad (2.37)$$

Performing the integration results in:

$$\left(-\frac{q \cdot F_s(u_s)}{k \cdot T}\right)^2 = 2 \cdot \lambda_i^{-2} \left[\cosh u_s - \cosh u_B - (u_B - u_s) \sinh u_B\right] \quad (2.38)$$

The mathematically and physically meaningful solution for the surface electric field $F_s(u_s)$ can be found by introducing the correct signum function $\text{sgn}(u_B - u_s)$ ([83], p. 56):

$$F_s(u_s) = \text{sgn}(u_B - u_s) (2)^{1/2} \frac{k \cdot T}{q \cdot \lambda_i} \left[(u_B - u_s) \sinh u_B + (\cosh u_B - \cosh u_s)\right]^{1/2}$$

$$(2.39)$$

CHAPTER 2. 1D MOS THEORY

In principle, for the calculation of the total capacitance C vs. the gate voltage V_G from F_s a formula is required that transforms gate voltage values into u_s values. This is *not* possible due to the interdependence of the band bending and the surface charge density Q_s. However, it is possible to go the other way round by using *fixed* band bending values and calculate the corresponding gate voltage values. To get total capacitance C vs. the gate voltage V_G curves, perform the following steps:

- With equation 2.24, 2.9 and 2.7 transform the "independent" variable u_s of the function $F_s(u_s)$ (equation 2.39) to obtain a function $F_s(\psi_s)$.
- Calculate $F_s(\psi_s)$ for different numerical values of ψ_s.
- Calculate $Q_s(\psi_s)$ with Gauss' Law 2.21.
- Calculate $C_s(\psi_s)$ with equation 2.22.
- If the oxide capacitance C_{ox} is known (equation 2.1), transform the chosen ψ_s values into gate voltage values V_G with equation 2.23.
- Finally, calculate the total capacitance C with equation 2.3, which works both with capacitance values ($\overline{C}$, $\overline{C}_s$ and $\overline{C}_{ox}$ in units Farad) as well as with capacitance values per unit area (C, C_s and C_{ox} in units Farad/cm^2).

2.3.4 Doping Level and Oxide Thickness Dependence

The most important and useful property of C(V) curves is their dependence on the various dielectric and semiconductor properties, hence the widespread applications of C(V) measurements and SCM as an analytical tool in material science. Chapter 3.4 gives an overview over *all* factors relevant in this work that influence the contrast in SCM measurements. This section is only dedicated to the influence of the doping level and the oxide thickness on the shape of the C(V) curves.

Figure 2.7 (a) shows a comparison of C(V) curves simulated for different doping levels. Increasing the doping level leads to an increase of the minimum capacitance value, because the depletion layer width w (equation 2.15) decreases with increasing doping level. Especially important for SCM, which records the first derivative dC/dV of the C(V) curve (refer to chapter 3.1), is that the C(V) curve's steepness at the transition from accumulation to depletion decreases with increasing doping level. Higher doping levels therefore lead to a reduced SCM signal. This is a way to measure doping level differences with SCM.

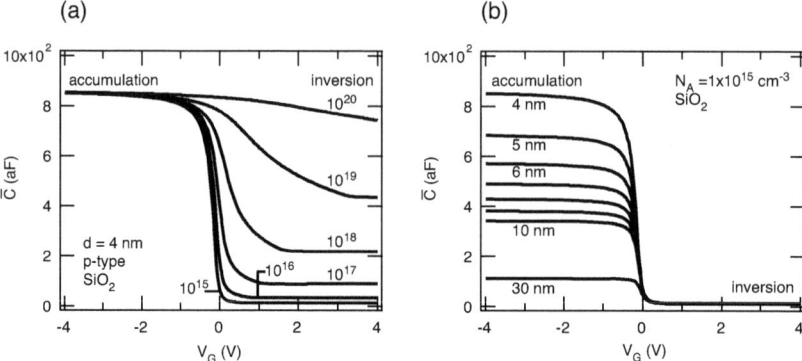

Figure 2.7: (a) Simulated C(V) curves with 4 nm of SiO_2 as dielectric on p-type Si with doping levels ranging from $N_A=1 \times 10^{15}$ cm^{-3} to $N_A=1 \times 10^{20}$ cm^{-3}. The oxide (accumulation) capacitance approximately remains constant, whereas the silicon (inversion) capacitance is dependent on the doping level. (b) Simulated C(V) curves with Si doping level of $N_A=1 \times 10^{15}$ cm^{-3} and a SiO_2 thickness ranging from $d=4$ nm to $d=10$ nm, and $d=30$ nm. The oxide (accumulation) capacitance is dependent on the oxide thickness d, whereas the silicon (inversion) capacitance remains constant.

For low doped samples, the semiconductor capacitance $C_s = \overline{C}_s/A$ in the inversion regime is nearly zero and due to equation 2.3, also the total capacitance $C = \overline{C}/A$ is nearly zero. Therefore, in the case of low doped samples, it is possible to derive the oxide capacitance from a measured C(V) curve even if the measurement setup suffers from large unpredictable stray capacitance which is common in capacitance measurements performed with an AFM tip. Because the stray capacitance is voltage independent, every capacitance value of an experimental C(V) curve is offset by the constant stray capacitance. For low doped samples, the stray capacitance value can be approximated by the minimum capacitance of the C(V) curve $\overline{C}_{min}$. The oxide capacitance is approximately the difference between the maximum capacitance $\overline{C}_{max}$ and the minimum. The relations are for low doped samples and are summarized in equation 2.40 and 2.41:

$$\overline{C}_{stray} \approx \overline{C}_{min} \qquad (2.40)$$
$$\overline{C}_{ox} \approx \overline{C}_{max} - \overline{C}_{min} \qquad (2.41)$$

For higher doped samples, the minimum of the ideal, simulated C(V) curve in figure 2.7 (a) is significantly different from zero. Therefore, in the case stray capacitance values are present, it is impossible to obtain the oxide capacitance by the measured C(V) curve alone.

Figure 2.7 (b) shows the dependence on the oxide thickness on the C(V) curves. Thinner oxides give a more pronounced capacitance difference between accumulation $\overline{C}_{max}$ and depletion $\overline{C}_{min}$. To be accurate, it is the ratio between the oxide thickness and the dielectric constant of the oxide, d/ε_{ox}, that is responsible for a pronounced shape of the C(V) curve (refer to equation 2.1). The lower this ratio, the larger is the oxide capacitance. To provide a good signal to noise ratio for the measurements, rather thin oxides with often large dielectric constants were used in this work, because this gives larger absolute differences in accumulation and depletion/inversion.

2.3.5 The Flatband Capacitance and Flatband Voltage

The easiest way to determine the flatband voltage of a measured C(V) curve is to calculate the total capacitance at flatband condition, $C_{FB} = C(V_{FB})$, for an ideal MOS system showing the same oxide thickness, dielectric constant and doping level as the measured sample. Taking the calculated flatband capacitance and looking for the corresponding gate voltage in the measured C(V) curve gives the flatband voltage of the measured curve. The semiconductor capacitance *per unit area* at flatbands, $C_{FBs} = C_s(V_{FB})$ is given by [83], page 84:

$$C_{FBs} = \frac{\varepsilon_0 \cdot \varepsilon_s}{\lambda_p} \qquad (2.42)$$

λ_p again is the extrinsic Debye length for a p-type semiconductor (equation 2.17). ε_s is the dielectric constant of the semiconductor. The total flatband capacitance, C_{FB}, can then be calculated using equation 2.3 and 2.42. C_{FB} therefore depends on the doping level as well as on the thickness and dielectric constant of the oxide layer. Figure 2.8 shows the behaviour of the normalized total capacitance at flatbands C_{FB}/C_{ox} ($=\overline{C}_{FB}/\overline{C}_{ox}$) at different doping levels and oxide properties. Note that for the low doped samples and oxide layer parameters used in this work, the normalized total flatband capacitance is between $C_{FB}/C_{ox} = 0.02$ and $C_{FB}/C_{ox} = 0.09$.

2.4 Oxide Charges

Until now, the oxide layer of a MOS structure was modelled to be an *ideal* oxide which is free of any charges that influence the behaviour of the MOS system and the shape of the C(V) curves. The equations for the total capacitance were only governed by the charges Q_s located on the semiconductor

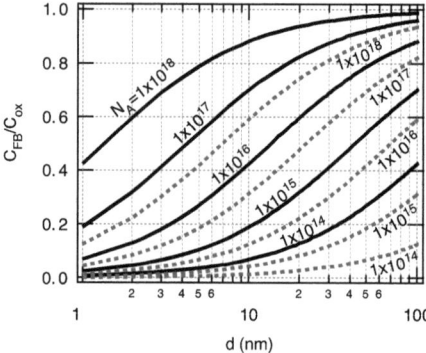

Figure 2.8: Normalized total flatband capacitance as a function of oxide thickness d with acceptor concentration as parameter at a temperature of 300K. The solid lines were calculated for SiO$_2$ (ε_{ox}=3.9) and the dashed lines are for ZrO$_2$ (ε_{ox}=20).

side of the oxide layer. However, real oxides suffer from various charges distributed throughout the oxide. Figure 2.9 shows the different types of charges that occur in SiO$_2$ ([109], page 380). In ZrO$_2$, which is also used in this work, some details in the oxide properties may be different from SiO$_2$. However, it is assumed that on a ZrO$_2$ covered sample a view monolayers of SiO$_2$ are in between the silicon sample and the ZrO$_2$ layer because of the tendency of Si to quickly form a thin *native* SiO$_2$ layer in ambient atmosphere. No measures were provided to avoid the build up of this native oxide layer prior to the deposition process [3]. Therefore it is assumed that figure 2.9 also gives a reasonable good approximation for the case of a ZrO$_2$ layer.

In the following discussion, all the various oxide charge densities illustrated in figure 2.9 are *areal charge densities* (charge in Coulomb per area, C/cm^2). There are four different types of charges found in SiO$_2$:

- **Interface trapped charges** Q_{it}: Between the mono-crystalline Si and the SiO$_2$ there is about one monolayer of non-stoichiometric SiO$_x$ (with $x < 2$). At the transition from the Si to the non-stoichiometric SiO$_x$ there exists a more or less large amount of dangling bonds due to the large difference in material properties of Si (crystalline) and SiO$_x$

---

[3]As described in chapter 4.2.1, the silicon samples were treated with HF to remove the native oxide layer prior the ZrO$_2$ deposition. However, only a few seconds in ambient air are enough for a partial regrowth of the native oxide. Therefore one can assume the existence of a thin native oxide layer between ZrO$_2$ and the silicon.

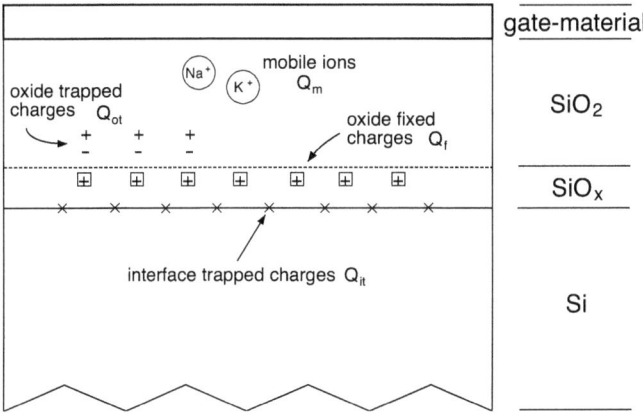

Figure 2.9: Different types of charges associated with thermally oxidized silicon.

(amorphous). These dangling bond give rise to the interface trapped charges Q_{it}. Because these traps have energy levels inside the band gap of silicon, and because they are located directly at the silicon surface, they can change occupancy with gate bias. They do not form energy bands.

- **Oxide fixed charges** Q_f: Further away from the silicon surface, there are the oxide fixed charges Q_f. They are immobile under applied gate voltage, and do not exchange charge with the silicon when gate voltage is varied (in contrast to the interface trapped charges). Furthermore, they are predominantly positive. It is assumed that they are located at a distance of 3 nm form the silicon surface ([109] page 390, [83] page 286). In electric measurements, Q_f can be regarded as a sheet of charge located at the Si-SiO$_2$ interface.

- **Oxide trapped charges** Q_{ot}: Inside the stoichiometric SiO$_2$ layer, there are the oxide trapped charges Q_{ot}. The controlled generation of these charges due to scattered hot electrons inside a MOSFET is used in electrically programmable read only memories (EPROM) [44].

- **Mobile ionic charges** Q_m: Finally, mobile ionic charges Q_m are mobile ions inside the oxide. They are mobile under electric fields at raised temperature. The main contributors are potassium K$^+$ and sodium Na$^+$ ions.

It is common to express the various oxide charges via the *number N* of charges per area (cm^{-2})

$$N = \frac{Q}{q} \quad (2.43)$$

In this way, Q_{it}, Q_f, Q_{ot}, and Q_m can be converted to N_{it}, N_f, N_{ot}, and N_m respectively. Note, that even if the charges are measured as areal charge densities, the impact on the MOS behaviour changes with varying areal density *as well as* with a redistribution of the charges inside the oxide, which leaves the areal density unaltered.

Concerning the reaction to a gate voltage change, there are two groups of charges. Q_f, Q_{ot} and Q_m do not respond to an applied gate voltage under normal conditions and only shift the C(V) curve on the voltage axis[4]. Only interface trapped charges Q_{it} can exchange charge with the silicon and therefore they are able to change their occupancy with the applied gate voltage. Therefore the interface charge density is dependent on the band bending $Q_{it} = Q_{it}(\psi_s)$. Because ψ_s is correlated to the gate voltage V_G via equation 2.23, changing the gate voltage also changes the occupation state of the interface traps. This leads to a broadening of the C(V) curve because the active amount of interface traps changes with voltage. Furthermore, in general, the interface trapped charges are not distributed evenly at the energy range of the band gap. Therefore, one can define the *interface trap level density* D_{it} (cm^{-2}eV^{-1}) ([83], page 308):

$$D_{it}(\phi_s) = -\frac{1}{q}\frac{dQ_{it}}{d\psi s} \quad (2.44)$$

An example where the interface trap density Q_{it} is not distributed evenly over the band gap is discussed in chapter 6.5.

The total impact of all charge types on the flatband condition, and therefore the flatband voltage is given by ([109], page 466)

$$V_{FB} = W_{ms} - \frac{Q_{it}(\psi_s = 0) + Q_f + Q_{ot} + Q_m}{C_{ox}} \quad (2.45)$$

Note the mutual dependence of the flatband voltage and the interface trap density. $Q_{it}(\psi_s = 0)$ can be translated to $Q_{it}(V_{FB})$ via equation 2.23. Therefore the value of the interface trapped charge density at flatband condition $Q_{it}(V_{FB})$ does influence the flatband voltage V_{FB} and vice versa.

It is possible to define a *total* oxide charge density at flatband condition $Q_{tot}(V_{FB})$:

$$Q_{tot}(V_{FB}) = Q_{it}(V_{FB}) + Q_f + Q_{ot} + Q_m \quad (2.46)$$

---

[4]There is some evidence that mobile ions do move under applied gate voltage also at room temperature. Refer to chapter 6.5.

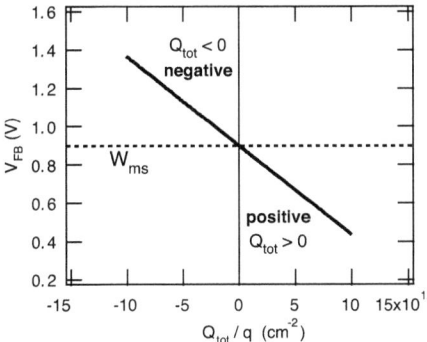

Figure 2.10: Flatband voltage V_{FB} vs. total oxide charge density Q_{tot} calculated from equations 2.45 and 2.46. W_{ms} is the work function difference between the diamond AFM tip and a Si-sample with an acceptor concentration $N_A = 9.4 \times 10^{14}$ cm^{-3}. The oxide capacitance $\overline{C}_{ox}$ was calculated using an oxide thickness of $d=5.12$ nm and a dielectric constant valid for ZrO$_2$ of $\varepsilon_{ox}=20$.

If the work function difference W_{ms} and the oxide capacitance C_{ox} is known, spacial fluctuation of the total oxide charge density $Q_{tot}(V_{FB})$ can be measured by determining the flatband voltage as described in chapter 2.3.5. Whether $Q_{tot}(V_{FB})$ is positive or negative can easily be determined by looking at figure 2.10. All measured flatband voltages that lie *above* the line defined by the work function correspond to a *negative* total oxide charge density and measured flatband voltages that lie *below* correspond to a *positive* total oxide charge density. With equation 2.45, it is possible to obtain the *sign* of the total oxide charge density measured via the flatband voltage by knowing only the work function difference W_{ms}. For a quantitative analysis, it is also important to know the oxide capacitance or the thickness and dielectric constant of the oxide.

2.5 Edge Effects – Deviations from the Flat Capacitor Model

Until now, the behaviour of the MOS capacitor was calculated neglecting any effects from fringing fields at the edge of the capacitor. These effects are commonly referred to as *edge effects*. However, for capacitor areas as small as an AFM tip this edge effects can not be neglected anymore. Correlated with the problem of fringing fields is the impact of the tip shape on C(V) curves. These are well known problems, and some theoretical and experimental work has been published that try to explain the impact of the special AFM tip shape on SCM measurements. Publications showing the effect of the tip shape and tip area that were found relevant by the author of this work, were done by D.M. Schaadt and E. T. Yu in 2002 [99] and G.H. Buh et al. in 2003 [20]. In their calculations they both showed that C(V) curves are broadened due to the very small area of the AFM tip, and that the broadening is increased if the tip area is further reduced. Besides the broadening of the C(V) curves, also the depletion layer width w (equation 2.15) decreases with decreasing tip area. This quantitatively changes the minimum capacitance $\overline{C}_{min} = C_{min} \times A$ of a measured C(V) curve and introduces an error in the determination of the stray capacitance. In addition, an error is also introduced if oxide capacitance $\overline{C}_{ox} = C_{ox} \times A$ is calculated from the difference of the minimum and maximum capacitance of a measured C(V) curve on a low doped sample (equations 2.40 and 2.41). Furthermore, the flatband voltage V_{FB} may change with decreasing tip area, introducing errors in all values derived from the flatband voltage.

However, to take the influence of a small tip area into account, the development of advanced simulation routines would have been necessary, which was beyond the scope of this experimental work. Therefore, edge effects were neglected for simplicity, and the unavoidable errors where accepted. Fortunately, the work of D.M. Schaadt et al. [99] indicates that all the errors associated with the small tip area are not severe above a limit of a (flat) tip area [5] of about 40×40 nm^2. It is difficult to compare a flat tip area and a tip area that is given by the *radius of curvature*, that assumes a spherical tip apex. However, it seems that the flat tip area limit of 40×40 nm^2 is at least in the range and probably below a flat tip area created by a spherical tip with a radius of curvature of about 100 nm, which is the value found in the data sheets of the used SCM tips. Furthermore, the limit is lower than the calculated electrical tip area of about 76×76 nm^2 for the tip used in chapter 6.6.

Therefore, in summary, one can say that the errors introduced by the edge effects due to the small AFM tip are probably not very dramatic if oxide capacitance and oxide charges are determined, even if the author of this thesis recommends to have a closer look into this subject in the future.

---

[5]The flat tip area is the area of the tip that is parallel to and in contact with the sample. It was calculated from the tip model presented in [99], taking a tip radius of curvature of 60 nm and a bluntness angle of 20 degree.

Chapter 3

Equipment

This chapter is dedicated to the description of the main equipment that was used for this work. In more detail, that is the scanning capacitance module of the multipurpose SPM and an ultra- precision (true) capacitance/loss bridge. Although the multipurpose SPM (Digital Instruments Veeco Dimension DI3100) was the basic component, it is not discussed, because its detailed functionality is of no relevance here. For an overview over SPM methods and applications refer to the introduction (chapter 5.1). When a detail of the SPM's functionality was relevant for a certain topic of this work, this special detail was described "locally" in the particular chapter. Efforts are made to explain as clear and as detailed as possible the difference between the *scanning capacitance module* and the *capacitance bridge*. It is very important to understand the different behaviour of the two setups. From now on, the procedure of measuring samples will be called *capacitance spectroscopy* or *Scanning Capacitance Spectroscopy* (SCS) in the case the scanning capacitance module is used, and *Quantitative Scanning Capacitance Spectroscopy* (QSCS) in the case the high precision capacitance bridge is used. The prefix "quantitative" suggests that the obtained data are real, calibrated capacitance values in Farad. Quantitative scanning capacitance spectroscopy therefore delivers C(V) curves, where a Farad-value is plotted against a voltage. In Scanning Capacitance Spectroscopy only a voltage proportional to the first derivative $(dC/dV)(V)$ of the C(V) curve is recorded, due to the applied lock-in amplification. This voltage will be called *SCM signal* or (dC/dV) *signal*. Therefore, the (dC/dV) signal is actually given in *Volt*, and not in *Farad/Volt*. In principle, the conversion of the SCM signal from units Volt to units Farad/Volt is possible by applying a simple factor of proportionality, but in reality this conversion factor changes from measurement to measurement and depends on many other factors not always known by the user.

Furthermore, a detailed description is given on how to connect the capacitance bridge to the SPM and how to establish beneficial conditions for reliable measurements with this setup. A short essay about how to establish a sufficiently good electrical contact to the sample is given in chapter 3.3. Finally, in chapter 3.4, a summary is given about all the factors influencing the capacitance signal both in conventional SCM as well as in QSCS.

3.1 The Standard Commercial Scanning Capacitance Equipment

The Scanning Capacitance Microscopy (SCM) module is an optional extension of the DI 3100 multipurpose SPM. With the SCM module it is possible to quickly image the dC/dV value between the AFM tip and the sample while the tip is scanned over the sample. In this way, "capacitance information" of a specific sample region can be obtained. Refer to chapter 3.1.2 for more information about the SCM image mode. It is also possible to record $(dC/dV)(V)$ spectra on single sample locations (refer to chapter 3.1.1). Due to the very small AFM tip area ($< 100 \times 100$ nm^2), the tip-sample capacitance values are in the range of only 1 to 500 aF, which makes fast measurements a challenging task. To enable fast and very sensitive capacitance measurements, the SCM module was designed on the basis of an ultra high frequency (UHF) resonant capacitance sensor. Originally developed by RCA (Radio Corporation of America), it was intended to be used inside the read-out electronics of early VideoDisc-systems [88]

In principle, the SCM module detects the change of the voltage amplitude of an excited LCR resonator circuit. The circuitry is shown in figure 3.1. A change of the sample capacitance $\overline{C}_{probe}$ leads to a change of the resonant frequency f_0 of the LCR circuit and, supposing a constant excitation frequency f_{excit} and voltage V'_{excit}, this subsequently leads to a change of the voltage amplitude V_{excit} between the tip and the sample in the LCR resonator circuit. V_{excit} is the voltage that probes an approximate value for the differential capacitance (equation 2.5) of the tip-sample system. The high frequency (915 MHz) excitation voltage V'_{excit} is inductively coupled into the LCR circuit via the inductance L$_2$. V'_{excit} is in the range of 5 V to 10 V ([15] page 47). The resonant frequency f_0 can be tuned manually to gain the maximum sensitivity by varying a DC voltage on a varactor diode, which changes its capacitance value $\overline{C}_{adj}$. The *voltage* on the varactor diode is commonly referred to as the *Cap. Sensor Frequency* (unit: V) and can be adjusted between 0V and 10V (refer to the SCM manual [3]). During measurements,

CHAPTER 3. EQUIPMENT   39

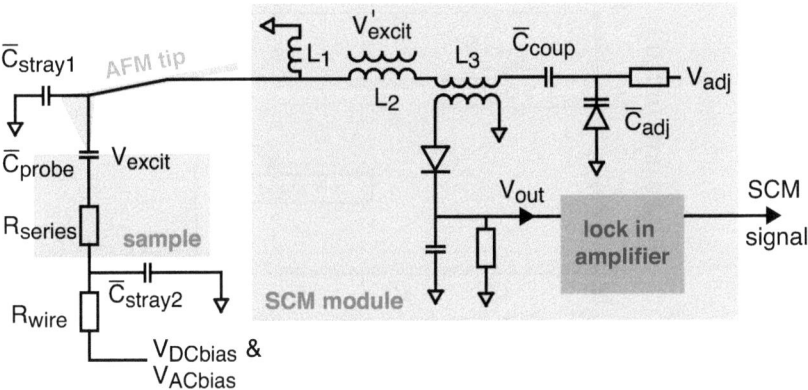

Figure 3.1: Circuit of the capacitance sensor inside the SCM module. The tip-sample system acts as an external branch of the sensor's resonant circuit. The resonator's amplitude is rectified and the resulting signal (V_{out}) is fed into a lock-in amplifier. The voltage output of the lock-in amplifier actually is the SCM signal proportional to the dC/dV signal. The schematics was taken from [15] and was modified afterwards.

the Cap. Sensor Frequency is held constant. The decoupling capacitor $\overline{C}_{coup}$ prevents any DC voltage contributions from the varactor diode circuit form entering the other parts of the resonator circuit. A DC bias voltage V_{DCbias} can be applied on the back-contact of the sample to change the potential between tip and sample. The tip voltage V_{tip} is simply given by

$$V_{tip} = -V_{DCbias} \qquad (3.1)$$

Due to the inductance L_1, there exists a DC short circuit to ground. For the sake of completeness, the series resistance inside the sample and the resistance of the feed cable is also included in figure 3.1. Via inductance L_3 the amplitude of the LCR resonator is fed into an rectifier where it is also smoothed. Every change to any of the capacitors in the LCR circuit leads to a change of the output voltage V_{out} of the rectifier circuit.

Because of the fact that *both* the stray capacitance $\overline{C}_{stray1}$ and $\overline{C}_{stray2}$ as well as the *change* of the stray capacitance values while scanning the AFM tip over the sample (in image mode) is orders of magnitudes larger than the tip-sample capacitance $\overline{C}_{probe}$, lock-in techniques have to be applied. Therefore, a second *AC bias* voltage V_{ACbias} is applied between the tip and the

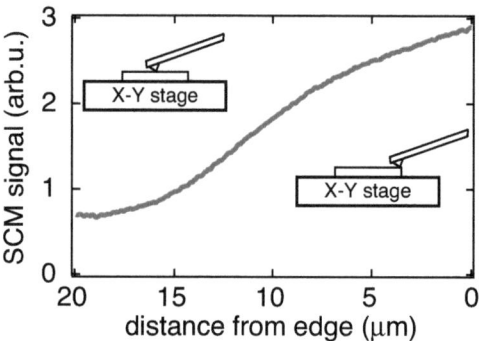

Figure 3.2: SCM signal vs. distance from sample edge. The SCM signal decreases steadily, because of the change (in this case the increase) of the stray capacitance between tip and sample.

sample that modulates the excitation voltage V_{excit}. The *AC bias frequency* can be adjusted between 10 kH and 170 kHz. The default value of 90 kHz was taken for all SCM measurements in this work. Due to the AC bias, the AFM tip alternately attracts and repulses the free carriers beneath the tip. The alternating carrier density under the tip may be modelled as a moving capacitor plate. Therefore, the signal obtained from the SCM module is not a capacitance value but the *change* of the capacitance with the applied AC bias: dC/dV. The applied "lock-in procedure" ensures that only capacitance changes are recorded that are due to the sample capacitance. Changes of the stray capacitance values due to the moving AFM tip are approximately filtered out. However, if large scan fields are chosen, the stray capacitance may vary significantly during a single scan line. Figure 3.2 shows the development of the SCM signal if the SPM tip moves from the edge of the sample to a location further away from the sample edge. There are two effects explaining this behaviour and this effects probably both contribute to the observed behaviour.

- One explanation is that the large *change* of the stray capacitance simply leads to a detuning of the resonator circuit. Subsequently, the sensitivity of the circuit decreases and the SCM signal decays. In principle, the strong impact of the change of the stray capacitance values on the sensitivity of the resonant circuit can be attenuated to a certain extent by varying the voltage on the varactor diode (Cap. Sensor Frequency). However, it is not possible to adjust the Cap. Sensor Frequency for every point of a scan line. Therefore, scanning over large areas can cause problems like shown in figure 3.2.

- The other explanation assumes the existence on an increasing severe short circuit for the high frequency ($f \approx 1$ GHz) voltage signal between the tip and the sample in the case the stray capacitance increases. The impedance of the stray capacitance is simply given by:

$$Z = \frac{1}{2 \cdot \pi \cdot f \cdot \overline{C}_{stray}} \qquad (3.2)$$

Assuming that the stray capacitance is about $\overline{C}_{stray}$=2pF...10pF, leads to a very low impedance of about Z=80Ω...16Ω, respectively. If one takes into account that high frequency voltage sources are commonly constructed for an impedance value of about 50Ω, the low impedance values of the stray capacitance will lead a to severe voltage drop.

A more detailed analysis of the impact of the stray capacitance on the SCM signal is a subject for future investigations. In general, recording small SCM images (e.g. 3×3 μm^2) near the sample edge is not affected by this problem, however it is recommended to adjust the Cap. Sensor Frequency in a way to obtain the largest sensitivity prior to the measurement.

The circuit inside the SCM module is sensitive to capacitance variations as small as aF (10^{-22}F/$\sqrt{Hz}$ sensitivity). For a more detailed description of the SCM sensor design refer to the SCM manual [3] or the original work from RCA [88]. For a detailed analysis of the SCM signal generation with a DI SCM module and a comparison to other capacitance sensors refer to the corresponding parts of the PhD thesis of Axel Born [15]. Unfortunately, the precise plan of the circuitry is not provided by Digital Instruments.

3.1.1 Spectroscopy Mode

The SCM module of the DI 3100 multipurpose SPM offers the possibility to record the differential capacitance dC/dV during a sweep of the DC bias voltage. However, as described in the previous chapter, the SCM module records the *change* of the capacitance, dC/dV, and therefore the the resulting curve is the first derivative of a C(V) curve. Nevertheless, in the following discussion this will be called a *capacitance spectrum*, and it will depend on the context whether dC/dV vs. V [(dC/dV)(V)] curves or C vs. V [C(V)] curves are meant. For better clarity, in figure 3.3 a simulated C(V) curve together with its first derivative (dC/dV)(V) is shown. It is important to

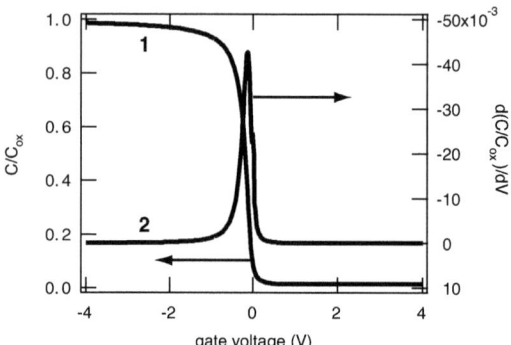

Figure 3.3: Comparison of a C(V) curve (1) and its corresponding dC/dV curve (2). Both curves were calculated assuming a flat, ideal capacitor with zero work function difference and no oxide charges (flat band conditions occurring at V_G=0V). The thickness of the SiO$_2$ layer was d=4nm and the Si sample had an acceptor concentration of N_A=1 × 10^{15} cm^{-3}.

note that the information content of a (dC/dV)(V) is less than that of a conventional C(V) curve. Lets first suppose ideal conditions without any stray-capacitance superimposed on the capacitance signal originating from the sample. As described in chapter 2, the oxide capacitance $\overline{C}_{ox} = C_{ox} \times A$ and the semiconductor capacitance $\overline{C}_s = C_s \times A$ (A... effective electrical tip area) in the depletion regime can be obtained from an ideal C(V) curve for all doping levels. However, the absolute value of the semiconductor capacitance in the depletion regime can not be determined in the case of a (dC/dV)(V) curve. Indeed, a C(V) curve can still be calculated from the measured (dC/dV)(V) curve by a simple numerical integration procedure, but the C(V) curve obtained in this way is determined except for an additive constant. Therefore it is *theoretically impossible* to determine the doping level of a semiconductor by a single measurement with the DI SCM sensor. Under certain circumstances, it is possible to determine unknown doping levels with multiple measurements, if there exists the possibility of performing calibration measurements. The very large stray capacitance values that happen under real measurement conditions mathematically *do not* affect the calculation of doping levels from multiple (dC/dV)(V) curves, because the stray capacitance only introduces another additive constant to the curve.[1]

---

[1]However, the large and unknown stray capacitance values *do* interfere with attempts to determine doping levels by measuring C(V) curves with the setup described in chapter

However, too large stray capacitance values affect the *sensitivity* of the SCM sensor by effectively short-circuit the excitation voltage as already described, and maybe by severely detuning the resonator circuit [2]. This leads to an reduction of the overall signal to noise ratio. Refer to chapter 3.4 for a summary of all factors influencing the signal in measurements with the conventional SCM equipment and measurement with the setup described in chapter 3.2.

As already discussed in chapter 2.3.5, the maximum of the $(dC/dV)(V)$ curve in figure 3.3 *is not* the flatband voltage of the MOS system created by the AFM tip, the dielectric layer and the semiconductor. The exact position of the flatband voltage depends on the dielectric and semiconductor properties. However, for most practical applications a shift of the $(dC/dV)(V)$ curve's peak can be interpreted as a similar shift of the flatband voltage even if there remains the possibility of effects that may affect the $(dC/dV)(V)$ curve's peak but not the flatband voltage (e.g. an unfortunate mixture of different charges inside or at the interface of the dielectric layer.)

Because the SCM module applies large probe voltages (around 1V) beyond the small signal region, the $(dC/dV)(V)$ spectra are averaged over the probe voltage and appear significantly broader than the simulated $(dC/dV)(V)$ curves. In figure 3.4 one can see the extend of the broadening of a measured $(dC/dV)(V)$ curve when compared to a calculated one. Even if most of the broadening can be attributed to the large probe voltages, the curves are also slightly affected by geometric effects due to the very small tip area (refer to chapter 2.5 for a more detailed analysis on the impact of the tip shape)

3.1.2 Imaging Mode

The main feature of the SCM sensor is that it features fast enough capacitance measurements to record dC/dV images of a sample region within a few minutes. For the imaging mode used for this work, no DC voltage bias is swept but is hold constant over the entire image. The contrast between different sample spots simply is created by the change of the position and shape of the $(dC/dV)(V)$ curve due to local varying sample properties like oxide thickness or the doping level. For a full discussion of all factors that determine the capacitance signal and therefore the contrast in SCM images, refer

---

3.2. There, similar to conventional SCM, if the additive constant (introduced by the stray capacitance) is not known, doping level calculations are impossible, and one has to perform calibration measurements, too.

[2]Until now, no systematic experiments have been performed investigating the interaction between the Cap. Sensor frequency parameter, different stray capacitance values and the sensitivity of the SCM sensor (= SCM signal height). This is an important subject of future investigations.

CHAPTER 3. EQUIPMENT   44

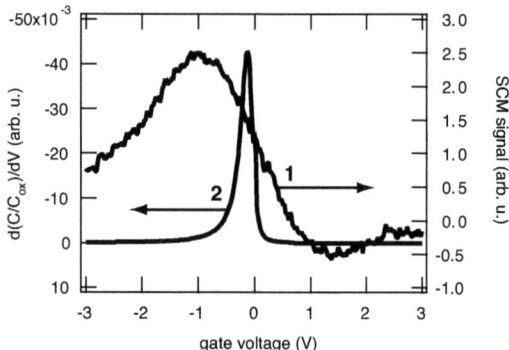

Figure 3.4: Comparison of a dC/dV spectrum recorded by the SCM module (1) and a simulated curve (2). Due to large excitation voltages applied by the SCM module, the measured curve (1) is significantly broadened. The measurement was performed on 4 nm of SiO_2 above a p-type silicon sample with an acceptor concentration of N_A=1 × 10^{15} cm^{-3}. The simulation was performed with the same parameters using the flat capacitor model and assuming flatband conditions at V_G=0V .

to the chapter 3.4. The constant DC bias can be chosen arbitrarily, either to gain the largest contrast on a specific sample, or to obtain a monotonic decreasing SCM signal for increasing doping levels. In any way, different doping levels always lead to different SCM signal and contrast, even if in most cases the SCM signal will **not** change monotonically with a monotonic change of the doping level (refer to chapter 3.5 or to the original article [105]). Figure 3.5 shows three SCM images of one and the same cross section of a bipolar transistor at different applied DC bias voltage values. The DC bias strongly affects the features of the transistor which can be observed in figure 3.5.

3.2 The Capacitance Bridge Setup

3.2.1 The Capacitance Bridge

The Andeen Hagerling AH2550A ultra-precision capacitance bridge used for this work operates at an excitation frequency of 1 kHz. It incorporates a true *bridge* circuit in the conventional sense of the word. The basic bridge circuit is shown in figure 3.6. The use of specially-wound ratio transformers and a temperature-controlled fused-silica capacitor standard in the basic bridge

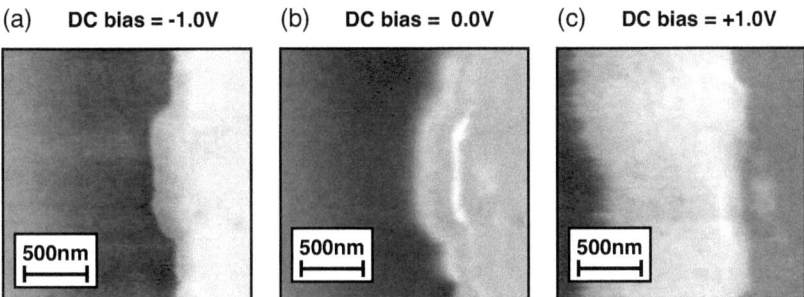

Figure 3.5: SCM images of one and the same cross section of a bipolar transistor recorded with different DC bias values V_{DCbias} $(= -V_{tip})$.

circuit are major contributors to the extremely high accuracy and precision that the AH2550A offers.

A 1 kHz sine wave generator feeds the ratio transformer which forms legs 1 and 2 of the basic bridge. Both of these legs have many transformer taps to allow selection of precisely defined voltages to drive legs 3 and 4 of the bridge. Leg 3 consists of one of several fused-silica capacitors plus other circuitry that simulates a very stable resistor. Leg 4 contains the unknown impedance, which is connected between "Low" and the "High". The microprocessor in the AH 2550A performs the task of selecting (or balancing) Taps 1 and 2 of the transformer and of selecting $\overline{C}_0$ and R_0 so that the voltage present at the detector is minimized. The detector is capable of detecting both in-phase and quadrature voltages with respect to the generator voltage. This allows both resistive and capacitive components of the unknown impedance to be independently balanced. In the case of this work, only the capacitance output was of interest, the resistive or loss output was only recorded to check for leakage currents.

If the microprocessor is able to obtain this null (or minimum voltage) condition during balancing, the unknown capacitance can then be determined since the ratio of the unknown capacitance $\overline{C}_x$ to $\overline{C}_0$ is equal to the ratio of the voltage on Tap 1 to the voltage on Tap 2. Similarly, the unknown resistance can be determined since the ratio of the unknown resistance R_x to R_0 is equal to the ratio of the voltage on Tap 2 to the voltage on Tap 1. The microprocessor performs these calculations and displays the capacitance and loss results.

Besides the high accuracy of the bridge, the most important feature is the possibility to input *maximum* values for the excitation voltage applied

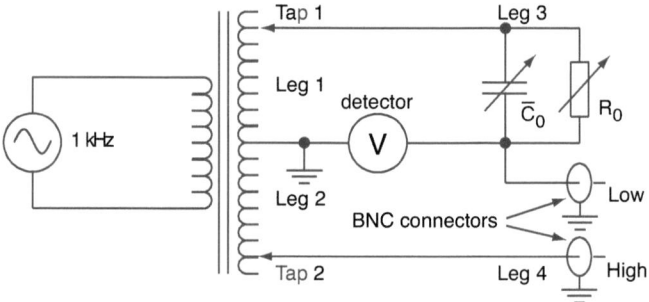

Figure 3.6: Basic circuit of the capacitance bridge. The device under test is connected to the "low" and "high" terminals.

between the "Low" and the "High" connectors as small as 1mV. This is the main advantage over the standard SCM equipment, where the applied excitation voltages can not be quantified by the user and where the voltages often cause problems because they are beyond the small signal region. Although the excitation voltages for measurements with the capacitance bridge were between 100 mV and 250 mV (chapter 6.2) to increase the signal to noise ratio at a manageable value, the applied voltages are nevertheless much smaller than in conventional SCM.

Here a summary of the important features of the AH 2550A capacitance bridge:

- adjustable excitation voltage between 0.001 V and 15 V
- measurement of calibrated capacitance values in Farad
- true resolution of 0.5 aF
- 8 significant digits of capacitance and loss measurements under optimal conditions
- independent measurement of capacitance and loss values
- high long term stability

3.2.2 Setup for AFM-Based Measurements with the Capacitance Bridge

Connecting a capacitance bridge to an AFM is not as straightforward as one might think initially. Whereas conventional SCM is not very sensitive to the influence of light, either from the ambient illumination or from the class 2 feedback laser[3], capacitance measurements with the (low frequency, 1kHz) capacitance bridge turned out to be *extremely* sensitive to even marginal illumination, e.g. from the microscope camera lighting. Especially the minority carrier density is severely increased by the influence of light, which subsequently has a large impact on the depletion and the inversion of a MOS system. This effect is well known to everyone who performs classical C(V) measurements on devices. Therefore, it is mandatory to switch off the feedback laser in order to obtain a useful capacitance signal. However, the requirement of switching off the feedback laser for the capacitance measurements leads to a series of necessary additional modifications to the setup, which will be described in this chapter. A scheme of the whole setup with all the additional features required for reliable capacitance measurements with the AH2550A capacitance bridge between the AFM tip and the sample is given in figure 3.7. The setup mainly consists of the SPM, the capacitance bridge, a circuit for switching on an off the feedback laser and a circuit for providing an external voltage to the z-piezo. In addition, a thermometer is required for precise temperature control. The two camera pictures below the scheme in figure 3.7 show the DI multipurpose SPM inside the acoustic hood (on the left side) and the SPM controllers and the circuit that provides the external z-piezo voltage (on the right side).

The AH2550A capacitance bridge does not provide a (non-zero) DC bias voltage to be applied to the sample via the "high" and the "low" lines (refer to chapter 3.2.1). However, a DC bias sweep is required to perform C(V) spectroscopy. For this purpose, the AH2550A bridge provides an input channel for an *external* DC bias in the range of ±100V. Therefore, a *Keithley 2400 source measure unit (SMU)* was connected to the bridge as a programmable voltage source and delivers a DC bias on demand. Inside the bridge, the

---

[3]Empirical data suggest that the conventional (high frequency, ≈1GHz) SCM signal is only increased by about 20% to 30% if the feedback laser is switched off. A reason why the conventional SCM signal shows only a very small sensitivity to light may be the very high frequency of the excitation voltage of about 1 GHz compared to the excitation frequency of the AH2550A capacitance bridge of only 1kHz. The light generated minority carriers may not be able to follow a modulation of about 1 GHz. Investigation on that topic may be a subject of future research. Two article showing some investigations on the impact of light on conventional SCM are given in [103, 20]

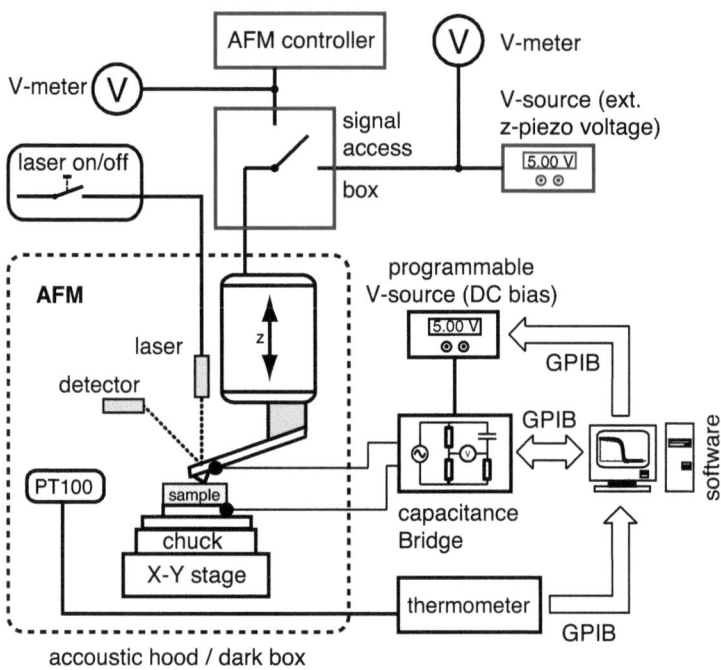

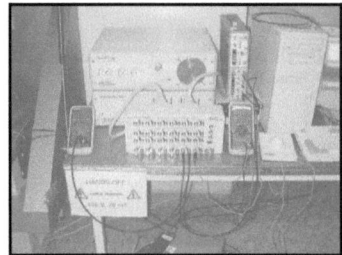

Figure 3.7: The experimental setup to perform SPM based measurements with the AH2550A capacitance bridge. Basic layout of the schematic by courtesy of Matthias Schramböck.

input channel carrying the (external) DC bias voltage is relayed [4] via one of two resistors to the "low" measurement line. According to the bridge's manual [1], chapter 4, page 13, choosing the resistor with the *higher resistivity* of 100MΩ results in measurements showing *reduced noise* compared to the outcome if the 1MΩ resistor were taken. The drawback of using the larger resistor, however, results from the larger $R \cdot C$ time constant of the circuit. Therefore, in principle, longer delay times have to be implemented between single measurements corresponding to the larger $R \cdot C$ time constant. Taking an approximate combined sample and stray capacitance value of 1pF and the larger 100MΩ resistor results in a time constant of about 0.1ms . This is even much shorter than the acquisition of a single capacitance measurement with the setup, and therefore the large 100MΩ resistor was taken for all measurements in this work.

Both the capacitance bridge as well as the DC voltage source (Keithley SMU) were controlled via a computer program written in *LabView*. The GPIB bus system was used for (bidirectional) communication. The program performed the task of recording C(V) curves and saving the data in text format on the hard disk. It automatically sweeps the DC bias voltage in the chosen range, performs delays between the single capacitance measurements and records a predefined number of C(V) curves and corresponding loss vs. voltage curves for averaging.

As already discussed, the influence of the feedback laser makes it impossible to do capacitance measurements with the bridge, and therefore it has to be switched off. Unfortunately, the AFM control software (*NanoScope 5.12r* was used for the work presented here) does not support switching the laser on and off. Therefore, the scanning head of the SPM was modified in a way to control the laser via an external switch. As it turned out, this was relatively simple because the SPM head features a security interlock system that switches off the laser if the whole scanning head is rotated from its operational vertical position to the horizontal position during handling. This was meant to reduce the probability of severe eye damage (class 2 laser) during handling of the scanning head. The main part of this security system was a position dependant mercury switch. It was very easy to bypass the mercury switch for the purpose of an external control of the laser. Note that the laser switch has to be located *outside* the acoustic hood, because opening the acoustic hood for the purpose of switching the laser on or off leads to severe

---

[4]Via a remote or front-panel command the capacitance bridge can toggle an internal relay between three possible states: either the DC bias input channel is *not* connected to the "low" line, or a connection to the "low" line is established via a 1MΩ *or* a 100MΩ resistor, optionally.

CHAPTER 3. EQUIPMENT

acoustic and thermal perturbations! Ultimately, this can result in tip- or sample damage or the separation of the tip from the sample. In both cases, no useful capacitance data can be obtained anymore. Even if the tip or the sample would not be damaged, additional time (at least 10 hours) would be required for thermal equilibration.

During the first test runs of the system containing only the SPM, the bridge with the external DC bias source and the laser switch, it quickly turned out that the SPM's feedback system reacted in a counterproductive way. If the laser is switched off, no laser light from the cantilever is reflected to the detector anymore. The feedback system, designed under the assumption that the laser always is switched on (when the tip is in contact with the sample), interprets a lost detector signal as a strong deflection of the cantilever. As a reaction to this supposed deflection, the feedback system tries to extend or retract the z-piezo in order to reduce the deflection of the cantilever. Finally this results in total extension or retraction of the z-piezo. No reaction of the feedback system can reflect the laser back on the detector's centre, because the laser simply was switched off. This again can lead to damage to the tip or the sample or a lost contact between tip and sample, which both makes reliable measurements impossible. Whether the z-piezo is extended or retracted may depend on the initial state of the cantilever's deflection at the time the laser is switched off. This subject was not investigated further.

To avoid the problems with an "overreaction" of the feedback system, an circuit providing an *external* voltage to the z-piezo was added to the setup, which can be seen in the upper part of the setup's scheme in figure 3.7. The central component of the external piezo voltage circuit is the *signal access box*. The signal access box allows the access to a large number of (analog) signals that are exchanged between the SPM controllers and the SPM, e.g. the x-, y- and z-piezo voltages. It provides the possibility of both monitoring the different signals as well as of feeding external signals into the system. In this way it is possible to quickly switch between the (automatic) SPM controllers and the external, manual piezo voltage circuit. A cheap, manually adjustable voltage source is connected to the signal access box. Currently, the voltage source only provides unipolar voltages in the range of 0 V to 100 V, but may be substituted in the future by a bidirectional voltage source with a possible voltage range between -100 V and +100 V.

Prior to switching off the feedback laser, the present z-piezo voltage applied by the SPM controllers is measured by a voltmeter (on the left side of the scheme in figure 3.7). Although the SPM control software also displays the voltage on the z-piezo, the external voltmeter was retained as a backup. After the read out of the current z-piezo voltage, the obtained voltage value is manually input into the external voltage source. Because the display of the voltage source is not very accurate, a second V-meter (the right one in the scheme of the setup in figure 3.7) was used to verify the output voltage of the source. Note that the z-piezo shows an expansion coefficient of about 14 nm/V; an accuracy of 1nm is obtained only if the external voltage can be adjusted with an accuracy of about 0.1 V. However, in the majority of cases the external voltage can differ from the voltage applied by the SPM controller by about 1V without having any effect. After adjusting the external voltage, the signal access box can be toggled to connect the external voltage to the z-piezo [5]. Finally, the feedback laser can be switched off. The current extension of the z-piezo is fixed by the voltage obtained from the external circuit. One also has to remember to switch off the illumination of the optical microscope inside the SPM. Capacitance measurements with the AH2550A bridge demand *complete* darkness!

Without any feedback the SPM can not adjust the pressure between tip and sample anymore. Therefore, every effect that leads to an deflection of the cantilever can not be compensated anymore. Scanning the tip over the sample (which is still possible because only the z-piezo remains fixed and the x- and y- piezo receive signals from the SPM controllers) is dangerous as long as the feedback is off both for the SPM tip as well as for the sample. It should be avoided except a *very* smooth and flat sample is investigated and the user knows what he/she is doing.

Especially thermal expansion can cause many troubles when the feedback is off. To get at least an estimate of the current temperature and its development inside the acoustic hood, a remote thermometer was deployed inside the acoustic hood consisting of a NTC resistor (PT100) and a multimeter to measure the temperature dependant resistance. The multimeter again was connected to a computer via a GPIB bus. Again a LabView program was used to convert the resistivity read out into Celsius temperature and to store the temperature values together with a time tag. For more information about the thermal expansion of the setup and temperature management refer to chapter 6.3.2. After the acoustic hood was open for a longer time, one has to wait at least one day to allow the system to thermally equilibrate again. Neglecting thermal equilibration lead to troubles with the tip-sample contact during measurements without feedback.

---

[5] Besides the z-piezo voltage, the signal access box also provides an the inverse $\bar{z}$-piezo voltage, with the opposite polarity to the z-piezo voltage. Because of the lack of detailed information on the $\bar{z}$ voltage and how it affects the functionality of the SPM, a third external voltage source (not shown in figure 3.7) was installed to control the $\bar{z}$ voltage. Therefore, prior to switching off the feedback laser, both the external voltages for the z-signal as well as for the $\bar{z}$-signal have to be adjusted properly. Variations to the $\bar{z}$-signal do not seem to affect the piezo, but maybe the SPM needs the $\bar{z}$-signal to work properly.

A very important issue when performing measurement in the aF regime is the way the device under test (in the case of this work the AFM tip and the sample), is connected to the bridge. The AH2550A capacitance bridge's manual strongly suggests to perform a proper shielding in order to avoid a reduction of the measurement sensitivity or strange measurement results. To cite a rule-of-thumb given in the bridge's manual: "If you see the object that is intended to be shielded, than it isn't!". Unfortunately, without major reconstruction of the SPM scanning head, it is not possible to perform a proper shielding. To keep the wording of the manual, "the tip-sample contact can be seen from outside". Figure 3.8 indicates the (non-ideal) way the AH2550A bridge is connected to the sample and the SPM tip. To connect the tip to the bridge, the conventional SCM tip holder was used. It features a wire that is electrically connected to the conductive AFM tip via clamps. In the case the conventional SCM is used, this wire is plugged into the SCM module. To connect the capacitance bridge to the wire (and consequently to the AFM tip), an electrical bypass was constructed and was glued to the SCM module. The bypass can be seen in the inset of figure 3.8. In this way a mechanically stable strain-relief was created that prevents the wire connected to the tip holder from experiencing too much traction from the relatively thick and stiff coaxial cable that is intended to connect the bridge with the AFM tip. The wire on the tip holder is soldered on the clamp that holds the tip. Too much force applied to the wire eventually could open the clamp and release the AFM tip. Note, that the lowest structure on the scanning head is the electrical bypass now instead of the AFM tip. Therefore one must take care not to accidentally crash the electrical bypass into the sample chuck or into the sample, which could damage the scanning head.

To support at least as much electrical shielding as possible, a coaxial cable is used to connect the "high" terminal of the bridge with the electrical bypass. Similarly, the "low" terminal of the bridge is connected to the sample holder. The construction of a sample holder turned out to be crucial, because the potential of the chuck interferes with the bridge measurement. In conventional SCM the chuck us used to apply the DC bias voltage, but it can not support this function in the case of the bridge setup. The sample holder simply consists of an insulating epoxy board, with a small copper structure on the top side. On one side of the structure the coaxial cable from the "low" terminal can be plugged in, and on the other side one can place the sample. To obtain a good electrical contact, the copper structure is gold plated. For more information about the appropriate back contact of the samples and how it can be established refer to chapter 3.3. To compensate for the strain applied to the sample holder by the stiff coaxial cable, the sample holder was

CHAPTER 3. EQUIPMENT 53

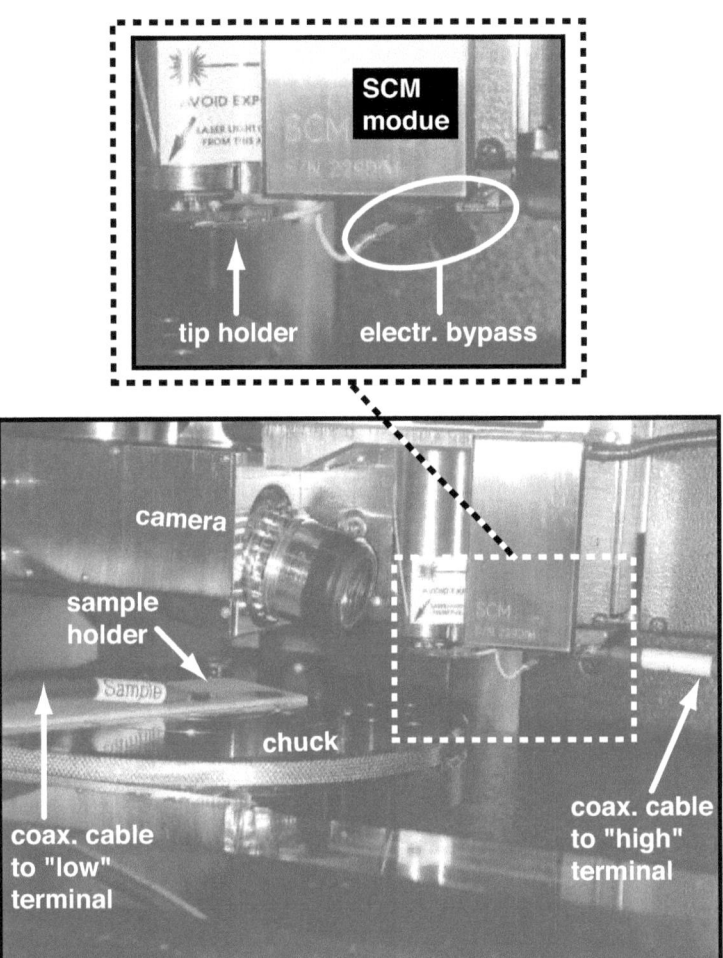

Figure 3.8: Images illustrating the way the cables plugged in to the "high" and "low" terminals of the capacitance bridge were connected to the SPM tip and the sample, respectively. The contact from the "low" cable to the sample was mediated by the sample holder. The "high" cable was connected to the tip via an electrical bypass. The inset shows a magnified image of the bypass.

fixed by a vacuum applied between the sample holder and the chuck.

The way the connection between the bridge and the tip-sample system is established is certainly far from being ideal, because of the unsatisfactory electrical shielding. Nevertheless, it turned out that reliable capacitance measurements are possible with this setup. The electrical noise that is generated inside the acoustic hood by the SPM electronic gives rise to a noise of about 60 aF during capacitance measurements with the bridge setup. [6].

3.3 Establishing an Electrical Back Contact to the Sample

A non-ideal back contact to the sample has different consequences in measurements performed with conventional SCM and with the new bridge setup. For example, nice, conventional SCM images rich in contrast can be recorded even on samples that are electrically isolated from the sample chuck. In contrast to that, it is impossible to record any C(V) curves with the bridge setup without a *very* good contact to the sample. The effort that is spent on establishing an appropriate back contact to a sample will depend on the given problem. For quick SCM images that are only intended to display qualitative changes in the lateral sample properties, it is not important to worry a lot about the back contact. One can simply place the sample on the sample chuck in an arbitrary way, with the only restriction that the resulting stray capacitance value is not too large. For the impact of the stray capacitance on conventional SCM refer to chapter 3.1. However, if spectroscopic measurements are required both using the conventional SCM [$(dC/dV)(V)$ curves] as well as using the bridge setup [C(V) curves] on single sample spots, a good electrical contact is needed to guarantee that the DC bias signal reaches the sample. In addition, a good contact is also required if conventional SCM images are recorded to gain quantitative information. For example, the quantitative analysis of doping gradients is only possible if the appropriate DC bias is present between the tip and the sample (refer to chapter 3.5 or the original article [105]). There are different ways how to establish a good electrical back contact to the sample. For imaging applications with the SCM module, a sufficiently good contact often is established by simply putting the sample on the chuck or a conductive spacer to avoid large stray capacitance values. A better contact can be established for cross sectional samples, if

---

[6]The noise of the data recorded by the capacitance bridge strongly depends on some internal parameters (average level, etc.) which can be chosen by the user. The parameter set which was most often used for this work, and which gives rise to 60 aF of noise, is described in chapter 6.2.

Figure 3.9: Cross sectional sample holder for SCM applications. The screw is used to fix the sample and to establish a good electrical contact.

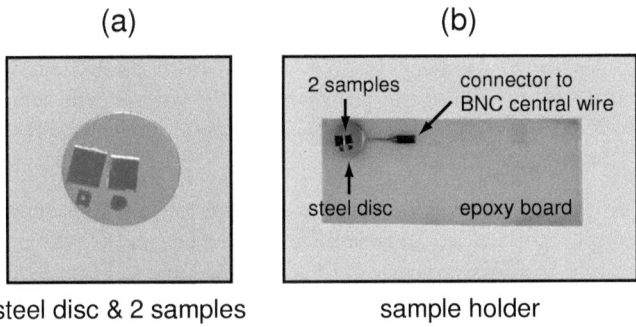

Figure 3.10: (a) Samples that are glued to a gold covered steel disc with conductive silver. (b) Disc with samples put on the sampleholder for measurements with the capacitance bridge setup.

the samples are screwed on the cross sectional sample holder 3.9. There, the metallic screw is pressed hard against the sample for a good contact. High quality back contacts can be obtained by covering the back side of the (silicon) sample with sputtered aluminium and a subsequent annealing step between 300 °C and 400 °C. Sebastian Golka reported variations of the SCM signal depending on the applied back contact in his diploma thesis [34].

In this work, a sufficiently good back contact was established by using conductive silver and performing the following steps: First, the (clean) backside of the Si-sample was gently scratched with a glass cutter in order to remove the native oxide locally. Then, conductive silver was put on the scratched backside and the sample was glued to a small, gold plated steel disc. This can be seen in figure 3.10 (a). The disc can be placed on the (also gold plated) sample holder (figure 3.10 (b)), which creates a very good electrical contact. The sample was glued to the steel disc and not directly to the sample holder because this enabled quick sample changes, if other samples glued to steel discs are available. To overcome the adherence of a sample that was glued to a surface is quite tricky and finally may damage the sample. To avoid any damage, the samples were glued to removable discs.

3.4 Factors Influencing the Capacitance Signal

This part is intended to summarize all the factors that influence the capacitance signal both in conventional SCM as well as in measurements with the bridge setup (QSCS). Again the following relation is valid to transform DC bias voltage to tip voltage: $V_{DCbias} = -V_{tip}$. Both the influence of the experimental setup as well as the properties of the sample will change the capacitance signal, and these aspects will be discussed in the following paragraphs:

- **Tip Voltage:** The most important parameter for the MOS capacitance is the applied tip voltage. The voltage dependence on the capacitance is either given in C(V) curves or in $(dC/dV)(V)$ curves. Figure 2.6 shows the theoretical voltage behaviour of an ideal MOS system. Depending on the tip voltage value chosen to record for example an SCM image, different sample properties on different location of the sample will deliver a distinct SCM contrast. Refer to figure 3.5 to get an impression of the large impact of the tip voltage on the contrast in conventional SCM. Whereas the bridge setup always delivers *positive* capacitance values[7] regardless if p-type or n-type samples are investigated, the conventional SCM module may deliver negative dC/dV values for one doping type depending on the tip voltage[8]

- **Excitation Voltage:** To measure the differential capacitance of the tip-sample system, one has to apply a (small) AC excitation voltage V_{excit}. If V_{excit} is too large, the measured (nonlinear) C(V) curves will be smeared out because the single capacitance values will be averaged over the large applied V_{excit}. Unfortunately, in conventional SCM the

---

[7] Negative capacitance values may be delivered by the AH 2550A capacitance bridge, but this indicates either a connected inductance or an uncommon RC-circuit connected to the bridge. Refer to the AH 2550A bridge's manual [1], chapter 8, page 2.

[8] The algebraic sign of the conventional SCM signal (dC/dV) depend on the special operation mode (Phase Mode or Amplitude Mode) of the SCM module. For more information about the sign of the SCM signal and how to switch the operation mode refer to the SCM manual [3], page 22.

excitation voltage that is applied to the sample can be assumed to be quite large and, in addition, depend on the sample- and stray capacitance itself. Therefore, in conventional SCM the $(\mathrm{d}C/\mathrm{d}V)(V)$ curves are smeared out (refer to figure 3.4). There is no way to limit the excitation voltage in conventional SCM. However, excitation voltages that are applied by the AH 2550A capacitance bridge can be tuned manually by the input of arbitrary maximum values. Therefore the capacitance bridge setup enables measurements in the small signal regime, which prevents broadening of the measured C(V) curves. However, in case of conventional SCM as well as in the case of the capacitance bridge, the sensitivity (signal to noise ratio) depends on the excitation voltage. For optimal sensitivity, the excitation voltage has to be beyond the small signal regime in both cases. The reduced sensitivity of the bridge setup during measurements in the small signal regime can often be compensated by extensive averaging of the results.

- **AC Bias (conventional SCM only):** In conventional SCM, the height of the so called AC bias determines the height (and therefore the signal to noise ratio) of the $\mathrm{d}C/\mathrm{d}V$ signal (besides other parameters). The AC bias is used in the lock-in circuit of the SCM module to detect *changes* of the capacitance. The higher the AC bias voltage, the higher is the capacitance change and therefore the $\mathrm{d}C/\mathrm{d}V$ signal. However, large AC bias values also leads to a broadening of the $(\mathrm{d}C/\mathrm{d}V)(V)$ curves, because of averaging.

- **Cap. Sensor Frequency (conventional SCM only):** The counterintuitively named Cap. Sensor Frequency parameter is the *voltage* that is applied to the varactor diode inside the SCM module to tune the resonant frequency of the capacitance sensor. It is used to tune the capacitance sensor to highest sensitivity for changes of the tip-sample capacitor. Therefore, at optimum Cap. Sensor Frequency one obtains the highest peak in a $(\mathrm{d}C/\mathrm{d}V)(V)$ curve. Varying the Cap. Sensor Frequency always means to change the excitation voltage V_{excit} applied to the sample.

- **Doping Level:** A decreasing doping level will lead to an increase of the difference between accumulation and depletion as described in chapter 2.3.4. This also leads to an increase of the maximum of the absolute value of the SCM signal in the SCM spectrum. Furthermore, as described in chapter 2.3.5, the flatband voltage depends on the doping level. Therefore it is possible to detect different doped regions as different bright regions in an SCM image. If the tip voltage for SCM imaging is chosen properly, it is possible to obtain an monotonic SCM contrast for a monotonic change of the doping level (refer to chapter 3.5 or the original article [105]).

- **Oxide Thickness:** A decreased oxide thickness leads to an increase of the difference between the accumulation capacitance and the depletion capacitance (refer to chapter 2.3.4). The flatband voltage of the MOS system is shifted again in this case (chapter 2.3.5). Similar to the reduction of the doping level, this also increases the conventional SCM signal dC/dV.

- **Oxide Charges:** All types of oxide charges have an impact on the flatband voltage of a MOS system. Refer to chapter 2.4 for further details. In this work, oxide charge properties were investigated exclusively by the bridge setup, because well controlled, small probe voltage could be applied. Especially details of the interface trapped charge energy distribution could only be resolved by the bridge setup. Conventional SCM applies too large probe voltages, and therefore could not resolve such features (chapter 6.5). In principle however, it should be possible to investigate large flatband voltage shifts due to very large local changes in the oxide charge density by conventional SCM. SCM images where the tip voltage is held constant throughout the recording process should show a different SCM signal on sample regions of different oxide charge content.

- **AFM Tip Area:** An aspect unique to SPM based capacitance measurements is the effect of a changing tip-sample contact area. Taking equations 2.1 and 2.2 into account, a larger AFM tip area A always leads to an increased oxide and semiconductor capacitance. Especially if capacitance spectra, C(V) as well as $(dC/dV)(V)$, are recorded on different sample spots for comparison, a sudden change in the AFM tip area (e.g. due to tip damage or adhesion of dust particles) is very annoying because it renders the measurement series unusable. The same is true for conventional SCM imaging.

- **Stray Capacitance:** The stray capacitance is a very large capacitance value (from 100 fF to 10pF) parallel to the capacitance signal of the investigated sample spot. It consists of the (not shielded) wires that are connected to the sample and to the AFM tip. The main contributor, however, comes from the (parasitic) capacitance between the AFM cantilever and its projected area on the sample. Note that this is *not* the desired tip-sample capacitance. The larger the overlap between the

AFM cantilever and the sample, e.g. during measurements on sample spots far away from the sample edge, the larger is the stray capacitance. The impact of the stray capacitance on the conventional SCM signal measured with the DI3100 SCM module was already discussed in chapter 3.1. C(V) curves recorded by the bridge setup only show an additive offset corresponding to the stray capacitance value.

3.5 Mechanism of Bias Dependent Contrast in Scanning Capacitance Microscopy Images

The text and the figures presented in this chapter 3.5 are an excerpt from the original article [105] written by J. Smoliner, B. Basnar, S. Golka, E. Gornik, B. Löffler, M. Schatzmayr and H. Enichlmair in 2001. Despite the possibility to refer to the original article [105], an excerpt of the relevant part of the article was included in this thesis for reference purposes and because it completes the discussion of factors that influence the capacitance signal of conventional SCM measurements and measurements with the capacitance bridge (chapter 3.4). The subject of the following paragraphs was found of great importance for future SCM based investigations of doping profiles.

"[...] In this work, we investigate the physical processes leading to SCM contrast. Using conventional Metal-Oxide-Semiconductor (MOS) theory [109, 83], and an epitaxial staircase structure we show that the maximum SCM signal strongly depends both on doping and on the applied bias. In general it is found that SCM images can be ambiguous, since different doping concentrations can yield the same signal size. Only in accumulation, where the SCM signal decreases exponentially with doping concentration, and in depletion, where the contrast is reversed, unambiguous results are obtained.

The sample we used was a CVD-grown doping staircase prepared by AMS (Austria Mikro Systeme International AG) which consists of five nominally 400 nm thick p-type Si-layers with doping concentrations of 2×10^{15} cm^{-3}, 2×10^{16} cm^{-3}, 1×10^{17} cm^{-3}, 2×10^{18} cm^{-3} and 9×10^{18} cm^{-3}, respectively. The highest concentration is located at the sample surface. The p-doped silicon substrate has a concentration below 1×10^{15} cm^{-3}. The dopant concentrations were determined by Secondary Ion Mass Spectrometry (SIMS) measurements, the results of which are shown in figure 3.11.

To avoid the usual problems related to sawing and polishing procedures for cross-sectional AFM/SCM measurements the samples were cleaved and

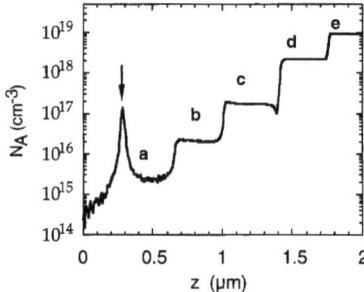

Figure 3.11: Doping profile of our epitaxial staircase structure determined by SIMS. The sample surface is on the right hand side. The peak at z=0.25 µm is an unintentional artefact of the epitaxial process.

subsequently oxidized in UV-light [48, 110, 119]. For the back contact, sputtered aluminium was employed. The capacitance measurements were performed using the Dimension-3100 system with integrated SCM sensor (Digital Instruments, USA). The probes implemented for the investigations were conductive diamond tips (Nanonsensors, Germany) which turned out to be superior to metal coated tips due to their high resistance against abrasion. Space charge effects in such tips can be neglected as long as the dopant concentration in the tip (1×10^{20} cm^{-3}) is much higher than in the sample.

Before we discuss our SCM data, we make a convention concerning the bias polarity: In analogy to textbooks on MOS theory [109, 83], the bias in this work is always plotted in a way as if it would be applied to the AFM-tip. In reality this is not the case, since in the DI-3100 SCM the bias is applied to the substrate for technical reasons. Furthermore one has to keep in mind that the SCM only measures the derivative of the capacitance, dC/dV, and not the capacitance itself.

Figure 3.12 shows cross sectional SCM images of our sample measured at different bias values. The sample surface is on the right hand side. Figure 3.12(a) was measured at a tip- bias of -1.9V and figure 3.12(b) at V=+0.8V. As one can see, the contrast between these two images is reversed. Figure 3.12(c) shows cross sections of SCM images, measured at four different bias values. Two features are evident: First, the 400 nm wide differently doped layers are clearly visible as well defined steps in the SCM signal (curve (1)). As a consequence we conclude that geometry effects of the tip can be neglected, otherwise the steps would be washed out. This washout, however is nicely seen for the doping spike at the substrate interface, the position of

CHAPTER 3. EQUIPMENT 61

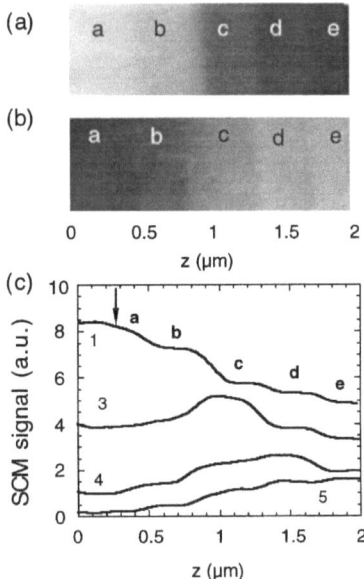

Figure 3.12: (a) SCM image of our sample taken at a bias of −1.9V. Image (b) was recorded at +0.8V. The regions (a-e) have doping concentrations 2×10^{15} cm^{-3}, 2×10^{16} cm^{-3}, 1×10^{17} cm^{-3}, 2×10^{18} cm^{-3} and 9×10^{18} cm^{-3}, respectively. (c) Sections through SCM images taken at a bias of +0.8V, 0V, −0.5V and −1.9V (curves 1, 3, 4, 5, respectively). The numbering of the curves corresponds to the numbering in figure 3.13.

which is marked by an arrow both in curve (1) and the SIMS data (figure 3.11). As the spike is much narrower than the steps and already in the same order as the radius of the tip (100 nm), only a small dip is observed instead of the expected well pronounced minimum.

As a second feature in figure 3.12(c), the contrast dependence as a function of bias can be seen in detail. At −1.9V (curve (1)), the SCM signal decreases with increasing doping. At +0.8V (curve (5)), however, this behaviour is reversed and the SCM signal increases monotonically with increasing doping concentration. For bias values of -0.5V and 0V the behaviour is non monotonic and the maximum of the SCM signal is observed in region (c) and (d), respectively.

Although the bias induced contrast reversal was already reported in the literature [98], a detailed study of this behaviour was not carried out up to now. To explain the origin of this behaviour, we consider an ideal p-Si/SiO$_2$/Al junction as model system and use conventional MOS theory. An ideal MOS structure consists of two capacitors in series, one is represented by the SiO$_2$ layer, the other by the depleted space charge region in the silicon underneath. Two extreme cases can be distinguished in such a structure. At strongly negative bias ("accumulation"), holes from the p-silicon are attracted to the oxide. In this case, the width of the space charge region is zero and the capacitance becomes constant, since it is only determined by the oxide capacitance. For positive bias, the holes are repelled from the oxide and the width of the depleted space charge region increases. Now, the decreasing capacitance of the space charge region becomes dominant. The total capacitance decreases as well and slowly saturates at low values ("depletion"). In general the total capacitance exhibits a complex behaviour as a function of bias, doping and other parameters. For details concerning this topic the reader is referred to the books by Sze [109] and Nicollian and Brews [83].

Figure 3.13 (a) illustrates the influence of the acceptor concentration in Si on the calculated C(V) curves. Curve (1-5) were calculated for acceptor concentrations of 1×10^{15} cm^{-3}, 1×10^{16} cm^{-3}, 1×10^{17} cm^{-3}, 1×10^{18} cm^{-3}, and 1×10^{19} cm^{-3}, respectively. For the calculation an oxide thickness of 1.5nm (a typical thickness for native SiO$_2$ frequently used in SCM) and no traps or surface charges were assumed.

As one can see in figure 3.13 (a), the C(V) curves are stretched and shifted if the acceptor concentration is increased. The corresponding dC/dV curves are shown in figure 3.13 (b). Here the influence of doping manifests itself as a shift of the dC/dV-peak positions to more positive bias while the peak amplitude decreases. This behaviour already explains the origin of the non-monotonic behaviour of the SCM contrast. Let us assume that an SCM image is recorded in the accumulation regime at bias V1 (see figure 3.13 (b)) and that the tip is moved continuously from a low doped region ($N_A = 1 \times 10^{15}$ cm^{-3}) into a very high doped region ($N_A = 1 \times 10^{19}$ cm^{-3}). By this procedure, the corresponding peak in the dC/dV curves is shifted to the right from low to higher bias values. Since V1 is already on the left hand side of the peak (1), the dC/dV signal will decrease monotonically. In contrast, at bias position V4 this is not true because V4 is located on the right hand side of the peak (1) position. While the dC/dV peak approaches position V4 from the left in figure 3.13 (b), the SCM signal will increase with doping. At V4, the dC/dV signal will have a maximum and beyond V4, the dC/dV signal will decrease. Only if the bias is large enough, so that the sample is driven sufficiently deep into depletion (V5), the peak will not reach position V5 and therefore dC/dV increases over the whole range of doping concentrations.

Quantitatively, this behaviour is seen nicely if dC/dV is calculated as a function of N_A taking the sample bias as parameter. As our considerations apply for n- and p-type samples and the sign of the SCM output signal depends on the phase adjustment of the built in lock-in amplifier, we consider the absolute values of dC/dV for convenience. Figure 3.13 (c) shows the result of our calculation. The curve labels correspond to the bias values marked by arrows in figure 3.13 (b). If the sample is sufficiently deep in accumulation (curve V1) the SCM signal decreases exponentially with increasing doping. For more positive bias (V2), the SCM signal shows a clear maximum for a doping concentration around $N_A = 1 \times 10^{16}$ cm^{-3}. This maximum shifts to higher concentrations when the sample is driven deeper into depletion (curves V3, V4). In addition, the signal size decreases. At a bias of V5 the dC/dV maximum is already beyond $N_A = 1 \times 10^{19}$ cm^{-3}. Now the SCM signal increases monotonically in the whole regime between $N_A = 1 \times 10^{15}$ cm^{-3} and $N_A = 1 \times 10^{19}$ cm^{-3}, but it is much smaller than at bias position V1. Note that the bias, where $(dC/dV)(N_A)$ exhibits a maximum, is not identical with the peak position in $(dC/dV)(V)$. The $dC/dV(N_A)$ maximum occurs at somewhat higher bias than the peak in $(dC/dV)(V)$ since the peaks shift and decrease simultaneously with increasing doping.

To verify our model experimentally, we measured dC/dV curves with our SCM. Figure 3.13 (d) shows typical data obtained on the low doped substrate. Again, the absolute value of the SCM signal is plotted for convenience. Compared to the calculated dC/dV curve of an ideal p-Si/SiO$_2$/Al junction (see figure 3.13(b)), the position of the peak is shifted to negative bias, which is probably due to surface charges and the use of a diamond tip having a different surface barrier height than aluminium. In addition, the peak is much broader, which is mainly due to the tip geometry [*L.Ciampolini, private communication of J. Smoliner et al.*]. The arrows (1,3,4,5) indicate the bias positions, where curves (1,3,4,5) of figure 3.12 (c) were measured. Bias position (1), which is on the left of the maximum in the measured dC/dV curve, corresponds to bias position V1 in figure 3.13 (b), which is on the left hand side of the peak in the dC/dV curve (1). As predicted by theory, a monotonically decreasing SCM signal is observed with increasing doping level. The other bias positions (3,4,5) in figure 3.13 (d) can also be identified with the bias positions (V3,V4,V5) of our model. At bias position (3) and (4) the SCM signal shows a maximum for the layers having doping concentrations of 1×10^{17} cm^{-3} and 2×10^{18} cm^{-3}, respectively. At bias position (5) a monotonic decrease of the SCM signal is observed in our experiment. Although the agreement between our measurements and the simple model is surprisingly good and also the reproducibility of the SCM measurements has considerably improved due to the use of conducting diamond tips [12], there is still a number of open problems: First, systematic measurements of dC/dV curves on the differently doped layers of our test sample have shown that the detailed shape of the dC/dV curves as a function of doping cannot be explained quantitatively in terms of primitive MOS models. Most probably, the assumption of a p-Si/SiO$_2$/Al junction is a too crude approximation to describe the properties of the conductive diamond tip. Second, care should be taken on samples, where both p-type and n-type regions exist. If the bias is adjusted in a way that the sample is in accumulation in the p-type regions, it will be in depletion in the n-type regions. As a consequence, the contrast is reversed in the n-type region and also the signal will be small. Moreover, the contrast behaviour in the vicinity of pn-junctions will probably not be predictable by simple models. [...]"

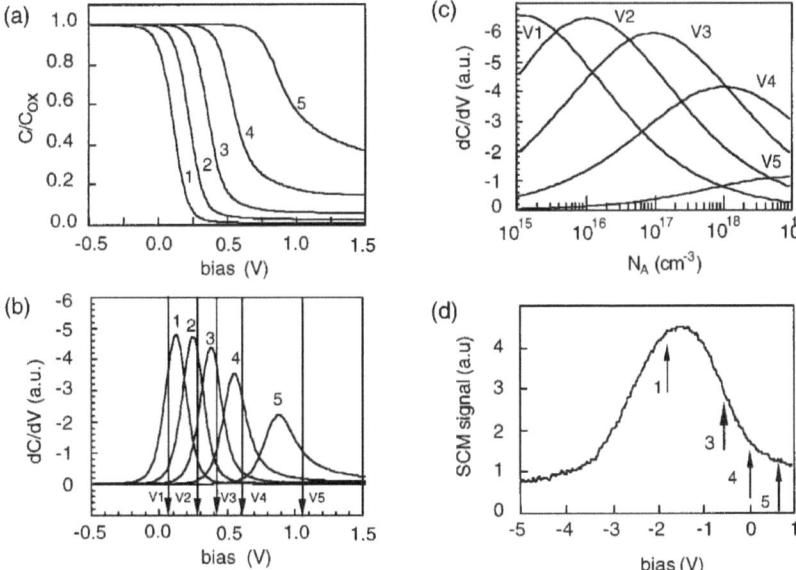

Figure 3.13: (a) Calculated C(V) curves of an ideal p-Si/SiO$_2$/Al junction. Curves (1-5) were calculated for acceptor concentrations of 1×10^{15} cm^{-3}, 1×10^{16} cm^{-3}, 1×10^{17} cm^{-3}, 1×10^{18} cm^{-3} and 1×10^{19} cm^{-3}, respectively. (b) Corresponding dC/dV curves. The y-axis of the dC/dV plot was flipped for convenience. (c) dC/dV plotted as a function of N_A for different bias values as labelled with (V1-V5) in figure (b). (d) Typical dC/dV curve measured with our SCM. The arrows labelled with (1,3,4,5) indicate those bias values, at which curves (1,3,4,5) in figure 3.12 (c) were taken and correspond to the bias positions (V1,V3,V4,V5) in our calculation.

Chapter 4

ZrO$_2$ as Dielectric Material for Scanning Capacitance Microscopy

4.1 Introduction

In this chapter, ZrO$_2$ is introduced as a high quality dielectric for SCM. SCM is not an easy and straightforward technique to use. Quantitative reproducible measurements are a serious problem, since the preparation of the insulator required on the sample surface (up to now exclusively SiO$_2$) has a dramatic influence on the results, especially in cross sectional measurements. Native SiO$_2$, which forms within seconds on top of a cleaving edge after the Si sample is cleaved in ambient air is only a bad choice for SCM due to its unfavourable oxide properties. Too many oxide charges and a bad SiO$_2$-Si interface lead to unstable SCM signal conditions [16]. On the other hand, standard high temperature industrial oxidation yields excellent oxide quality, but cannot be used because on processed devices, the very high oxidation temperature (900 °C to 1200 °C) broadens all doping profiles or destroys the samples completely. Thus, special low temperature oxidation processes have to be used. As reported in literature, reasonably good low temperature SiO$_2$ layers are obtained on samples polished with silica slurry [33, 51], followed by a low temperature oxidation in an oven [19, 100]. Alternatively, irradiation with UV light and simultaneous oxidation through in-situ generated ozone [48], or a combination of these two approaches [110, 119], can be employed. Another possibility which yields better oxide quality than native oxide is wet chemical oxidation using liquid oxidisers [16]. Although low temperature oxidation processes are good enough in many cases where just qualitative

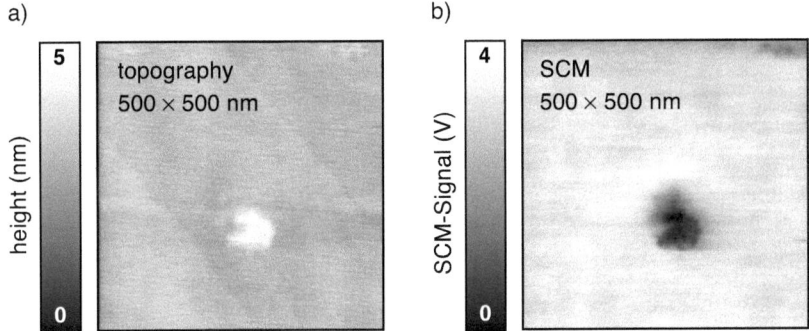

Figure 4.1: Impact of electrochemical oxidation on the topography and the SCM signal. a) The height of the hump near the centre is about 1.2 nm. b) The SCM signal at the location of the hump is less than 10% of the value without electrochemical oxidation.

SCM signals are required, there is still an urgent need for a reliable, low temperature deposition process for high quality insulators. For this reason, the SiO$_2$ layer was replaced by a high quality, low temperature chemical vapour deposition grown ZrO$_2$ layer. Since ZrO$_2$ is a high-ε dielectric, with an effective dielectric constant of thin and ultra thin films in the range of ε_{ox} = 20 [70], the same oxide capacitance can be achieved with a much thicker oxide layer (refer to equation 2.1). In comparison, SiO$_2$ only yields a dielectric constant of ε_{ox} = 3.9 [109]. To simplify the comparison of dielectrics with different dielectric constants, the concept of the *equivalent physical oxide thickness* (EOT) was introduced. The EOT is defined as the thickness of the SiO$_2$ layer inside a capacitor showing the same capacitance as another capacitor made out of a different dielectric material and having the same area. Two capacitors with different dielectrics show the same EOT if they have the same dielectric constant to thickness ratio ε_{ox}/d. Therefore, SiO$_2$ and ZrO$_2$ show the same EOT if the ZrO$_2$ layer is 5 times thicker than the SiO$_2$ layer.

Replacing SiO$_2$ with ZrO$_2$ (showing an increased thickness) also eliminates the second very severe drawback of SCM on SiO$_2$ covered silicon samples: SCM signal degradation due to dielectric breakdown of the SiO$_2$ layer or degradation due to large parasitic leakage currents. Furthermore, tunnelling currents influence the SCM signal, if SiO$_2$ layers of only a few nm thickness are used. These parasitic effects are commonly referred to as *charging effects*.

If grown properly, thicker dielectric layers are much less susceptible to leakage currents and subsequent *electrochemical oxidation processes*. Such electrochemical oxidation sometimes occurs if the applied voltages in SCM are too large or the same sample spot is treated for a longer time. Figure 4.1 shows what happens after measuring many dC/dV vs. V spectra on a Si sample covered with a very thin SiO_2 layer. The topographic image (figure 4.1 (a)) shows a small hump due to electrochemical oxidation. Figure 4.1 (b) shows the SCM image with a dark spot near the centre corresponding to the hump in the topographic image. Inside the affected area, the SCM signal is reduced by at least 90%. The reduction of the SCM signal is due to two effects. The first reason for a reduced SCM signal is, according to equation 2.1, the decrease of the oxide capacitance $\overline{C}_{ox}$ due to an increase of the oxide thickness d. This increase happens because of electrochemical oxidation. The second reason is an accumulation of oxide charges at the affected area which results in a shift of the flatband voltage. This leads to a different SCM contrast even if the oxide thickness would remain the same. These severe effects can be avoided by using thicker high-ε dielectrics.

Various methods can be used for the preparation of ZrO_2 layers such as reactive sputtering at room temperature or elevated temperature [82], chemical vapour deposition (CVD) from inorganic precursors [92, 22], and CVD from metal-organic precursor substances (MOCVD) [9, 10, 35]. From these methods, only MOCVD fulfils the requirement of low temperature deposition.

Before exploring the possibility to replace the insulating SiO_2 layer required for SCM measurements by MOCVD grown ZrO_2, an introduction to high temperature oxidation and MOCVD is given in the following chapter.

4.2 Deposition of the Dielectric Layers

This chapter is intended to give only a brief description of the preparation of the dielectric layers used in the following chapters. For a more detailed investigation of the physics of the involved processes or the properties of the prepared dielectrics refer to original work by S. Harasek [37].

4.2.1 Cleaning Procedure Prior to Oxidation or Deposition

Prior to the oxidation or the deposition process, the samples were cleaned using a modified RCA cleaning procedure [55]. Organic as well as inorganic contaminants on the sample surface are removed by treating the samples first in oxidising alkaline and afterwards in oxidising acid environment. The cleaning is mechanically supported by performing the procedure in an ultrasonic bath and elevated temperature. For the exact conditions and chemical composition refer to [37].

Even at ambient temperature condition (300 K), a thin SiO_2 layer is created on Si surfaces exposed to ambient air. These *native* SiO_2 layer does not exceed a thickness of about 1 nm to 2 nm, and is of low quality concerning the amount of oxide charges. To get rid of this low quality native oxide, the samples are etched in hydrofluoric acid (HF, 38%) for a few seconds after the RCA cleaning. Afterwards the samples are quickly put into the oxidation or deposition oven, to prevent contamination and to minimize the regrowth of the native oxide.

4.2.2 High Temperature Oxidation Process for SiO_2

The state of the art technique to increase both the thickness as well as the quality of a SiO_2 layer on Si, is to apply very high temperatures during oxidation. Oxidisers are oxygen (O_2) in the case of *dry* oxidation as well as water vapour (H_2O) in the case of *wet* oxidation. Wet oxidation is preferred if thick layers of SiO_2 are required. The diffusion coefficient of O_2 in SiO_2 is very low. If a the SiO_2 layer has reached a certain thickness, the oxidation rate will become very low even at high temperatures. However, water molecules show a much higher diffusion coefficient inside the SiO_2 layer, which leads to an increased oxidation rate. The equations for oxidising Si are

$$Si + O_2 \rightarrow SiO_2 \qquad (4.1)$$

$$Si + 2H_2O \rightarrow SiO_2 + 2H_2 \qquad (4.2)$$

Whenever high quality SiO_2 layers were required for this work, a high temperature oxidation process was carried out in a standard industrial oxidation oven (PEO-601 from ATV Technologie GmbH, Germany) at 900 °C using a wet oxygen flow ($O_2 + H_2O$).

The growth of the SiO_2 layer uses silicon material and this leads to a movement of the Si-SiO_2 interface, which is schematically shown in figure 4.2. During an oxidation process resulting in a total oxide thickness of d, the Si-SiO_2 interface moves a distance of $0.44 \cdot d$.

4.2.3 Metal Organic Chemical Vapour Deposition of ZrO_2

Metal organic chemical vapour deposition (MOCVD) is a variant of chemical vapour deposition (CVD). In CVD one or two *precursor gases* are used

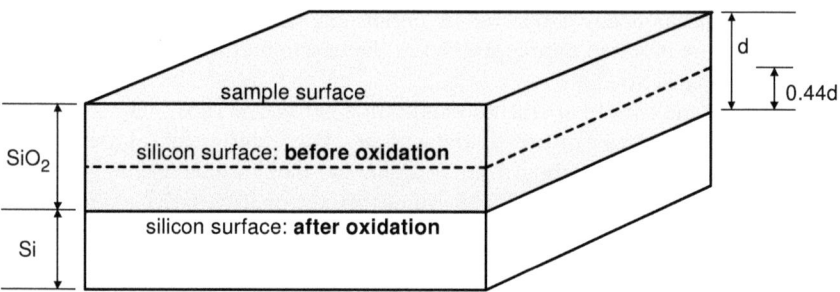

Figure 4.2: Scheme showing the movement of the silicon surface during oxidation. Prior to oxidation, the sample surface (ideally) equals the silicon surface. During oxidation, a silicon layer of thickness $0.44d$ is oxidised resulting in a SiO_2 layer of thickness d.

to create a layer of a solid material. After adsorption to the sample surface, either the two precursors react with each other, or only one precursor substance decays on the surface. In both cases a solid layer of the desired material is left on the sample and the gaseous by-products can be removed easily. Figure 4.3 shows a scheme of this process. The activation energy to start the reaction is either delivered by simply heating the sample, or, in case of plasma enhanced chemical vapour deposition (PECVD) [97], it comes from the plasma. Another possibility is to use the energy of ion [101, 76] or electron beams, e.g. from the focused ion beam system described in chapter 5.2. This enables very local deposition of materials.

Whereas CVD most often operates with *inorganic* substances at a temperature in between of 900 °C – 1200 °C [37, 92], MOCVD uses *organic* (carbon rich) precursors. These organic precursors are not as stable as their inorganic counterparts, and allow deposition temperatures as low as 350 °C. Unlike equations 4.1 and 4.2 which describe the oxidation of Si, deposition using organic precursors has no single and exactly defined material balance equation to describe the on going reactions. The reactions rather strongly depend on the actual conditions inside the reaction chamber (e.g. temperature).

In case of the ZrO_2 deposition, the precursor $Zr(tfacac)_4$ (zirconiumtrifluoroacetylacetonate) was used. It shows an advanced stability towards hydrolysis compared to other possible precursor materials [104, 90]. Whereas the inorganic precursors in CVD are most often gaseous at room temperature, the complex MOCVD precursors are most often liquid or even solid at

CHAPTER 4. ZIRCONIUM DIOXIDE AS DIELECTRIC FOR SCM

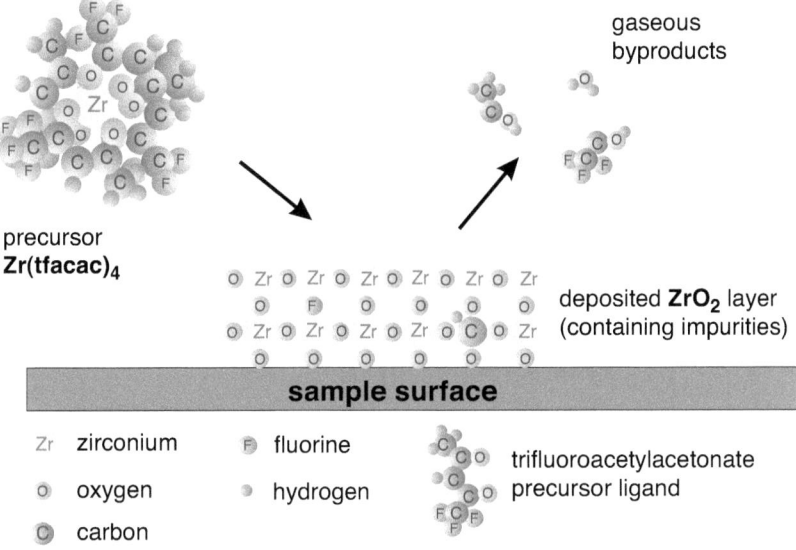

Figure 4.3: Scheme of the ZrO_2 deposition process. The precursor molecule adsorbs on the sample and decays into a ZrO_2 layer. Different gaseous by-products are created. Sometimes impurities are included into the ZrO_2 layer.

ambient conditions. Therefore, *bubbler systems* have to be used to deliver *gaseous* organic precursors to the reaction chamber. In figure 4.4 a scheme of the whole CVD system is shown. The bubbler system is heated to melt the precursor. Another reason why the precursor $Zr(tfacac)_4$ was chosen is because it shows a sufficient volatility even at moderate bubbler temperatures of about 140 °C. An inert carrier gas (argon) is blown through the molten precursor. If the temperature of the bubbler is right, the gas pressure of the liquid precursor is high enough to enrich the carrier gas. Finally it reaches the reaction chamber together with a certain amount of precursor gas. Note that a heated pipe has has to be used to connect the bubbler with the reaction chamber. Otherwise the precursor re-sublimates inside the pipes.

One can tune the amount of precursor gas which reaches the reactor by adjusting the bubbler temperature or the flow of the carrier gas (via the MFCs) through the bubbler system. Oxygen is added to the gas flow inside the reaction chamber, because it supports the deposition process. A second argon pipe is connected to steadily rinse the reaction chamber throughout

CHAPTER 4. ZIRCONIUM DIOXIDE AS DIELECTRIC FOR SCM   72

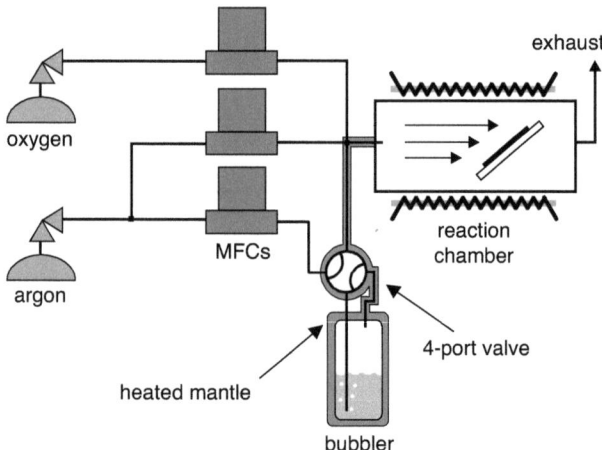

Figure 4.4: Scheme of the CVD system. The flow of the different gases are controlled via mass flow controllers (MFC). The bubbler system and all pipes containing the precursor gas are heated.

the deposition process.

The reaction chamber is a so called *hot wall reactor* operated at ambient pressure. It is heated electrically and can also be cooled actively. The samples have to be tilted to allow a homogeneous adsorption of the precursor from the gas flow.

Figure 4.5 shows the progression of a typical ZrO_2 deposition process. Prior to heating, the ambient atmosphere is removed from the reaction chamber by rinsing with argon. After reaching the appropriate deposition temperature and a short period for equilibration, oxygen and the carrier gas flow containing the precursor enter the reaction chamber. Note that the oxygen flow starts a short time before the precursor flow, in order to support the oxidising conditions ideal for deposition. Therefore, oxygen is also delivered after the end of the precursor flow, because the precursor concentration does not drop to zero suddenly. During cooling, only the argon flow is maintained.

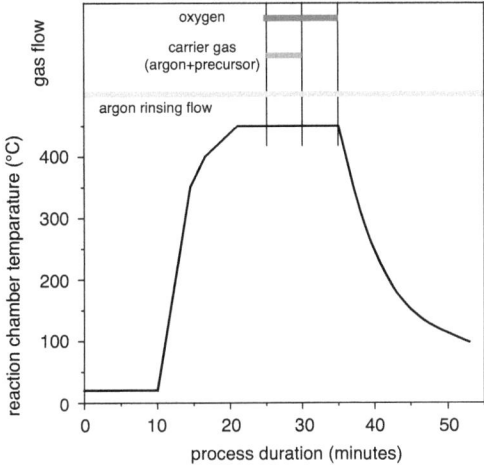

Figure 4.5: Typical temperature profile of the ZrO$_2$ deposition process. The corresponding gas flows of the different gases are given at the top of the figure.

4.3 Investigating the Properties of ZrO$_2$ as a High Quality Dielectric for SCM

To explore the electrical properties of ZrO$_2$ as a dielectric for SCM measurements, we used homogeneously p-doped Si samples with an acceptor concentration of N_A=9.4 × 10^{14} cm^{-3}. To show that ZrO$_2$ coatings can also be used to investigate more complex samples, SCM measurements were also done on ion implanted pn-junctions on p-type Si wafers. In contrast to the homogeneously doped samples, where measurements where performed simply on the wafer surface, the pn-junction sample had to be cleaved in order to uncover the junction. If done properly, cleaving yields a perfectly flat surface and avoids the usual problems related to sample sawing and polishing. To compare the standard high quality SiO$_2$ and MOCVD grown ZrO$_2$, the experiments where done with both types of dielectrics. All samples were subject to a standard RCA cleaning procedure [55] to get rid of organic and inorganic surface contamination followed by a treatment with HF to remove the very low quality native oxide. Efforts were made to reduce the time between native oxide removal and oxidation or deposition, to reduce the regrowth of native oxide to a minimum[1]. The high temperature oxidation process for the SiO_2 layers was carried out in a standard industrial oxidation oven (PEO-601 from ATV Technologie GmbH, Germany) at 900 °C using a wet oxygen flow as described in chapter 4.2.2. The deposition of the ZrO_2 layers was performed as described in chapter 4.2.3. To compare the quality of ZrO_2 and SiO_2 films in SCM measurements, the strong difference in dielectric constants has to be taken into account. While SiO_2 has a dielectric constant of $\varepsilon_{ox} = 3.9$, ZrO_2 is a high-ε material with a dielectric constant of $\varepsilon_{ox} = 20$. For a direct comparison we fabricated ZrO_2 films having the same EOT as the SiO_2 films. For the SiO_2 films we chose a thickness of 4 nm. The physical thickness of the corresponding ZrO_2 film was 20 nm, according to an EOT of 4nm.

4.3.1 Surface Roughness of ZrO_2 and Impact on the Resolution of SCM

In order to check the usability of ZrO_2 as a dielectric for SCM measurements, homogeneously p-doped silicon samples coated with ZrO_2 were investigated first. Figure 4.6 (a) shows a topographic surface scan of a ZrO_2 sample made in contact mode. When compared to the SiO_2 layer in figure 4.1 the ZrO_2 layer in 4.6 (a) shows a higher roughness. The arrows point to locations where the local thickness of the ZrO_2 layer is increased. This roughness has an impact on the SCM signal, which is shown in the simultaneously recorded SCM image in figure 4.6 (b). According to equation 2.1, an increased dielectric thickness d due to local topographic fluctuations translates into a local lowering of the SCM signal. Figure 4.6 (c) shows the topography and SCM signal along a line scan at nearly the same sample position. The correlation between increased topographic level and reduced capacitance signal is nicely visible. The ZrO_2 layer showed a surface roughness [2] of about 2 nm. The largest topographic features have a lateral extent W of about 30 nm, whereas features as small as approximately 10 nm are well reproduced by the SCM signal.

It is important here to understand that there are two relevant effects (among others, that are not important here, see chapter 3.4) that lead to a change

---

[1]In both cases of dielectric formation the HF treated silicon samples were exposed to the ambient atmosphere less than 5 minutes.

[2]This is the resulting surface roughness if diamond coated SCM tips (Nanosensors CDT-FM cantilever), with a tip apex radius of approximately 100 nm are used. Tapping mode measurements with Nanosensors NCHR cantilevers (tip apex radius about 10 nm) show an increased surface roughness of about 6nm.

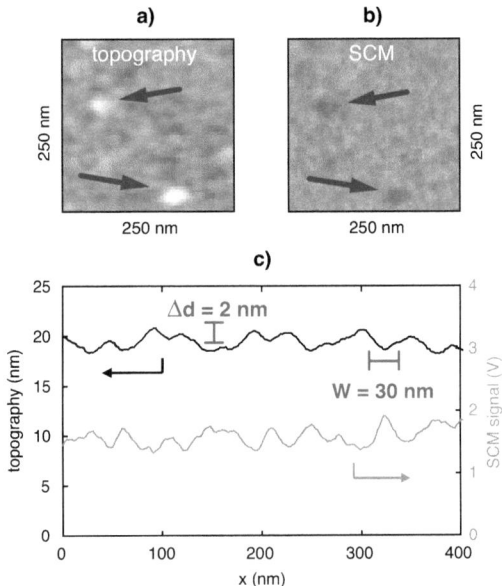

Figure 4.6: Correlation between ZrO$_2$ roughness and SCM signal. a) Recorded topography of the ZrO$_2$ layer in contact mode, b) simultaneously recorded SCM image of the ZrO$_2$ layer. Arrows point to locations where the local oxide thickness d is increased. c) A single scan line shows the strong correlation between the topography of the ZrO$_2$ layer and the SCM signal. The largest features have a height of about Δd=2 nm and a diameter of W=30 nm.

of the SCM signal. There is a difference between the topography induced changes of the SCM signal and the respond that is due to varying doping levels. There are two possible definitions of the lateral resolution of SCM, depending on the effect:

- First, one can investigate how the SCM signal changes at a sudden change of the oxide thickness. The SCM signal will *immediately* follow the topographic changes (figure 4.6 (c)). Therefore, oxide thickness induced changes of the SCM signal show the *same* resolution as the topography signal. Refer to chapter 2.3.4 more information about the influence of the oxide thickness on the capacitance signal.

- The second type of SCM contrast change is due to a varying doping level. Here the lateral resolution will depend on the applied AC and DC voltages and on the doping levels. Note that the flat capacitor model gives a maximum depletion layer width of about 1 μm at a doping level of 1×10^{15} cm^{-3} ([83], page 64). Furthermore, SCM is only sensitive to *charge carrier concentration* and not to acceptor or donator concentration. Therefore, for voltage conditions common in SCM, one can assume a spacial resolution of dopant profiling that is *much worse* than the topographic resolution of SCM.

For these reasons, SCM dopant profiling (e.g. investigations of pn-junctions) is not limited by the use of ZrO$_2$ showing a maximum lateral feature size of 30 nm and a roughness of 2 nm. In most practical cases SCM shows an effective lateral resolution for doping profiling that does not fall below 30 nm. Furthermore, for many applications the average over many scan lines can be taken to get the wanted information. Doing so, the topography induced capacitance fluctuations will cancel out in average without introducing other limitations into the measurement. Note that the deposition process may be optimized in the future to reduce layer roughness.

4.3.2 ZrO$_2$ Stability Against Electrical Stress

A big advantage in using ZrO$_2$ as a dielectric material for SCM turned out to be its high stability against electrical stress compared to the SiO$_2$ layers. As already explained in chapter 4.1, ZrO$_2$ is a high ε-material which makes it possible to use a much thicker ZrO$_2$ layer without a decrease of the oxide capacitance $\overline{C}_{ox}$. This increases the resistance against charging effects. To compare the ability to resist electrical stress, both a SiO$_2$ and a ZrO$_2$ sample having the same EOT were stressed by successively scanning a square of 500×500 nm^2 at a tip bias of V_{tip}=-3 V, as shown in figure 4.7. After 8 scans, the resulting oxide degradation was investigated by taking 5×5 μm^2 large images of the same area in topographic and capacitance mode. For the large scan V_{tip} was adjusted to yield the maximal signal in the corresponding unstressed areas, which was -1V on SiO$_2$ and +0.18V on ZrO$_2$ respectively. Figure 4.8 (a) shows the SCM image of the SiO$_2$ sample. In the unstressed region one obtained the expected signal size, whereas inside the stressed area the SCM signal was approximately zero[3]. This indicates a severe degradation of the SiO$_2$ due to charging effects while continuous scanning under high bias. A topographic line scan L of the stressed SiO$_2$ area is shown in figure 4.8 (b).

---

[3]Although the decrease of the SCM signal shows severe charging inside the treated are, the applied tip voltage and duration of the stressing procedure was not enough to induce electrochemical oxidation.

CHAPTER 4. ZIRCONIUM DIOXIDE AS DIELECTRIC FOR SCM

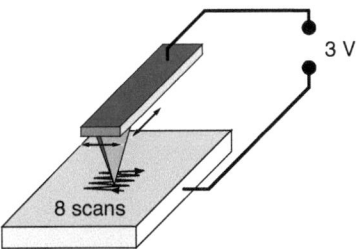

Figure 4.7: Setup for electrically stressing the samples. The conductive AFM tip was scanned 8 times over a area of 500×500 nm² at a tip voltage of 3V.

Apparently, no topographic anomaly was observed in the degraded region of the SiO_2 sample. As one can see in figure 4.8 (c), ZrO_2 turns out to be insensitive to electrical stress. Only a slight increase in the SCM signal is observed in the stressed region. Topographic investigations of the stressed region on the ZrO_2 layer show a reduced average thickness in the order of 1.5 nm, which is shown in figure 4.8 (d). Obviously, the ZrO_2 material was eroded by the AFM tip during the multiple scans, and as a consequence, the thickness of the dielectric material was reduced and an increased SCM signal was observed in this region. One can explain the larger abrasion of the ZrO_2 layer with a lower hardness of ZrO_2 compared to SiO_2.

4.3.3 Comparison of pn-Junctions Covered with SiO_2 and ZrO_2

To demonstrate the utilization of ZrO_2 as a dielectric material for SCM supported device analysis, SCM images of a pn-junction covered with SiO_2 and with ZrO_2 were compared. The pn-junctions were manufactured by ion implantation at Austria Mikro Systeme International AG (AMS) by implanting a highly antimony doped buried layer ($N_D > 1 \times 10^{20}$ cm^{-3}) into a low boron doped silicon wafer ($N_A < 1 \times 10^{15}$ cm^{-3}). The wafer had to be cleaved prior to the measurements, to expose the pn-junction. Before the comparison of the SCM images obtained with the different oxides, the tip-bias conditions, under which the images were taken have to be discussed briefly: A monotonic dependence of the SCM signal on the doping level is only achieved if the DC tip voltage is adjusted in a way that the SCM signal is *largest* in the lowest

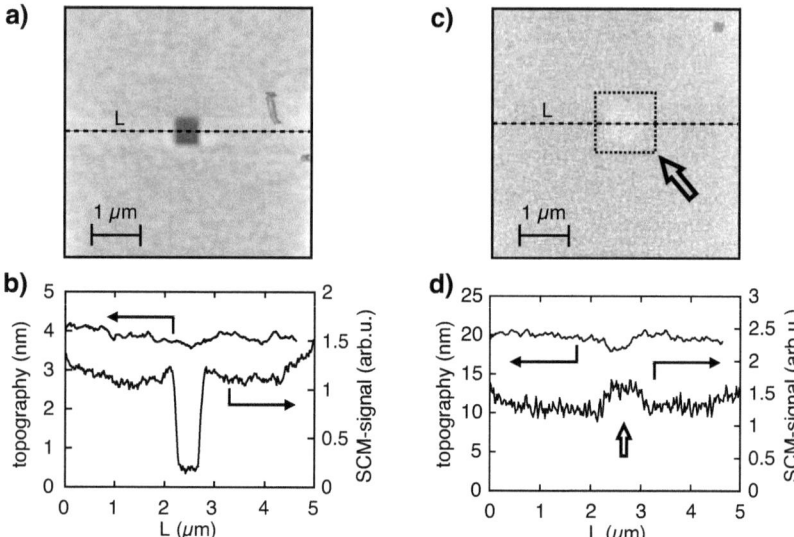

Figure 4.8: a) SCM image of the stressed region of a SiO_2 covered sample, the scan size is 5×5 μm^2. b) Topographic and SCM profile taken along line L of a). The charged area is clearly visible in the SCM signal, whereas the topography shows no obvious influence of stressing. c) SCM image of the stressed region of a ZrO_2 covered sample. d) Corresponding topographic and SCM profile of the ZrO_2 sample taken along line L of c). Only topographic lowering due to abrasion gives rise to a slight signal heightening.

doped region of a sample with doping gradients [4]. Adjusting the tip bias in this special way yields the largest SCM signal for the lowest doped regions (refer to chapter 3.5 or the original article [105]). Then, a decreasing SCM signal is obtained with increasing doping concentration. As a consequence, however, SCM measurements with these parameters in a region of doping with opposite polarity will result in a very low signal. However, as the focus here is on the properties of ZrO_2 for SCM measurements and not on a detailed pn-junction delineation [121, 29], the tip voltage V_{tip} was optimized at positive bias for maximum SCM signal in the p-doped bulk sample regions for simplicity. Adjusting the tip bias for maximum signal in the n-type areas,

---

[4]The DC voltage where the largest SCM signal occurs should not be mixed up with the flatband voltage on that particular sample spot. The position of the flatband voltage and the position of the highest slope of a C(V) curve are different! Refer to chapters 2.3.5.

CHAPTER 4. ZIRCONIUM DIOXIDE AS DIELECTRIC FOR SCM   79

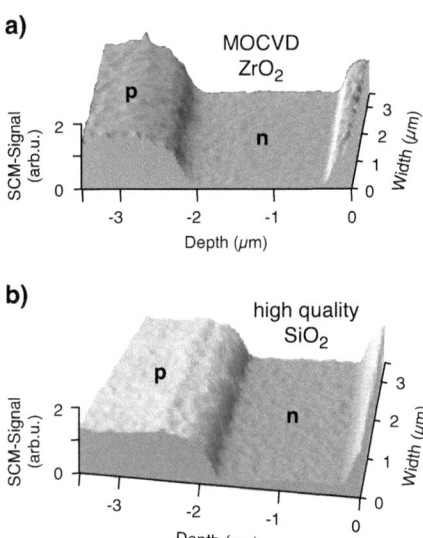

Figure 4.9: a) and b) SCM images taken on a ZrO_2 and a SiO_2 covered pn-junction, respectively. The pn-junction was exposed by cleaving. The wafer surface is on the right.

qualitatively yields the same results. As one can see in figure 4.9 (a), very good SCM contrast is obtained for the ion implanted region of a sample with a SiO_2 layer as dielectric material. The wafer surface is on the right hand side. The ion implanted highly n-doped region is visible as large valley in the 3D-plot of the SCM signal. The upper pn-junction is located directly underneath the sample surface and the lower pn-junction is observed in the region between -1.6 μm and -2.4 μm below the surface. If we compare these data with the results obtained from a ZrO_2 layer covered sample shown in figure 4.9 (b) (ZrO_2 layer having the same EOT as the SiO_2 layer), one can see that ZrO_2 can compete easily with the SiO_2 layers concerning contrast generation. Due to small, unwanted deviations from the tip bias which is needed to maximize the SCM signal in the p-doped bulk region on both samples, the depth profile of the SCM signal in the region of the pn-junction looks a little bit different on the SiO_2 and ZrO_2 sample. It is known, however, that the tip bias has a major influence on the carrier concentration in the vicinity of pn-junctions and consequently the measured junction-positions seen in figure

4.9 (a) and (b) can differ [29, 86, 87]. This also explains the slight differences in the width of the n-doped valley regions in figure 4.9 (a) and (b).

Chapter 5

Focussed Ion Beam Induced Damage Investigated by Scanning Capacitance Microscopy

5.1 Introduction

Focussed ion beam (FIB) techniques are among the most important tools for structuring of surfaces in the nanometre regime. Today, FIB systems are mainly used for device modification [102, 77], transmission electron microscopy (TEM) sample preparation [43, 115], scanning probe microscopy (SPM) tip preparation [63, 85], and deposition of different metals and insulators [76, 118]. However, there are also effects that limit the usage of FIB modification to certain applications and areal scales [62]. These effects are schematically displayed in figure 5.1. A very crucial factor for all FIB applications is the ion beam profile (figure 5.1 (a)). Difficult to measure, it defines the spatial resolution and the smallest possible size of a FIB made structure. One has to pay attention if FIB modification takes place near sensitive devices, which is illustrated in figure 5.1 (a). Literature indicates that the ion beam can be modelled consisting of two regions: the high intensity central beam and tails of very low intensity, which cover a much wider area [75, 7, 96].

Another important aspect is the amount of ions implanted into the sample and the related amorphisation and crystal damage[78], which can be seen in 5.1 (b). Depending on the acceleration voltage, the ions can penetrate deeply into the sample and scattering leads to a certain lateral straggling. Taking

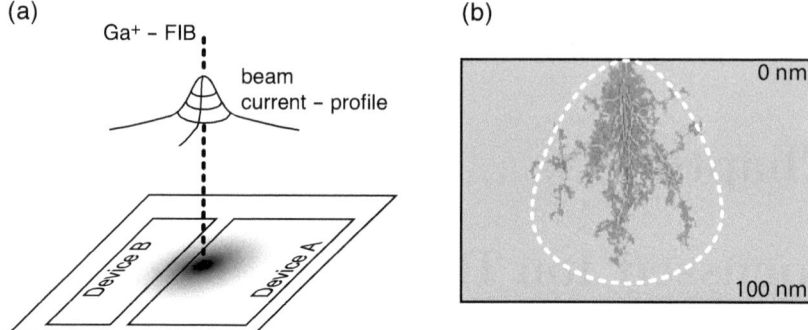

Figure 5.1: (a) The focussed ion beam profile consists of a high intensity interior surrounded by large beam tails which can do harm to devices near a location where FIB assisted sample modification takes place. (b) Simulation of the ion range and damage distribution inside a sample. The simulation was performed using the program *SRIM-1998*.

secondary effects such as recoils, channelling and end-of-range-defects into account, the disrupted sample area can be much wider than the actual beam area, and extend far into the sample. Various methods such as secondary ion mass spectroscopy (SIMS) and transmission electron microscopy (TEM) have been utilized to measure penetration depths and profiles of FIBs. However, the disadvantage of these methods are either their lack of 2D spatial resolution or difficult sample preparation. For this reason, scanning probe techniques have successfully been applied for FIB profile determination [69] and imaging in other ion beam applications[114]. Topographic atomic force microscopy (AFM) investigations can yield high resolution data from FIB implanted spots via the embossment of amorphisized areas due to the slightly lower density of amorphous silicon[23].

However, it is shown in this work that SCM can detect magnitudes smaller changes in material composition than any other method and it is possible to sense as small quantities as 10 - 100 impurity atoms per cubic micron (10^{13} - 10^{14} per cm^3). Where FIB-modification does not alter the topography of the region of interest, it is still possible to get reliable data of the FIB damaged areas via SCM. In chapter 5.4 SCM is first used to determine the shape of the ion beam. Chapter 5.5 finally deals with the damage depth and distribution of FIB modified areas.

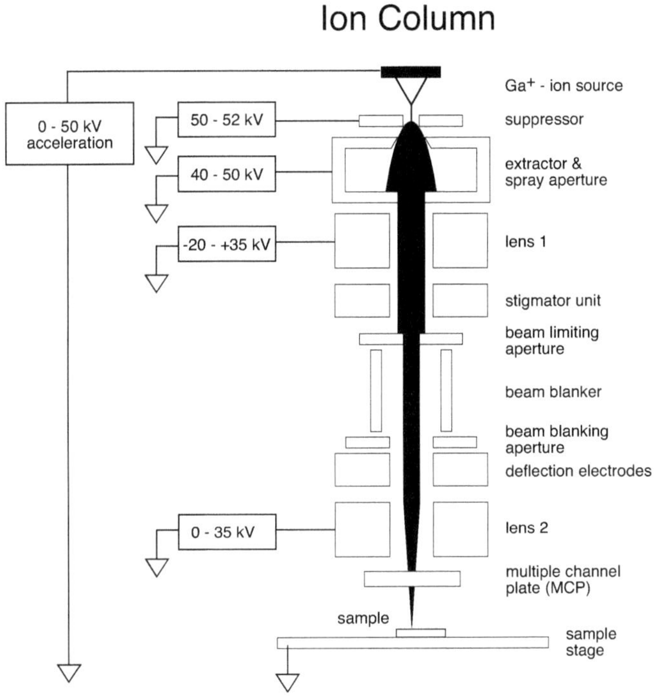

CHAPTER 5. FIB INDUCED DAMAGE INVESTIGATED BY SCM   83

Ion Column

Figure 5.2: Scheme of the ion column. The ion optics is based solely on electric fields. Magnetic fields are not applied in this design.

5.2 The Focussed Ion Beam System

Inside a focussed ion beam (FIB) system, ions are accelerated to high energies and then focussed to spot sizes of a few nm. The high energies and the very small beam diameter make it possible to perform arbitrary, local sputter-processes with a resolution below 100 nm [80, 73]. The ions are extracted and accelerated inside the *ion column*. Figure 5.2 shows an illustration. In the FIB system used for the following investigations Ga^+ ions are extracted from a so called *gallium liquid metal ion source* (LMIS)[1] by large electric

---

[1]The physical properties of Ga are very advantageous for applications inside ultra high vacuum (UHV) equipment: First, the very low melting point of 30°C requires only minimal heating and subsequent minimal heat transfer to the ion optics, which reduces the risk of misalignment. Second, Ga shows a very high boiling point of 2403 °C which

CHAPTER 5. FIB INDUCED DAMAGE INVESTIGATED BY SCM   84

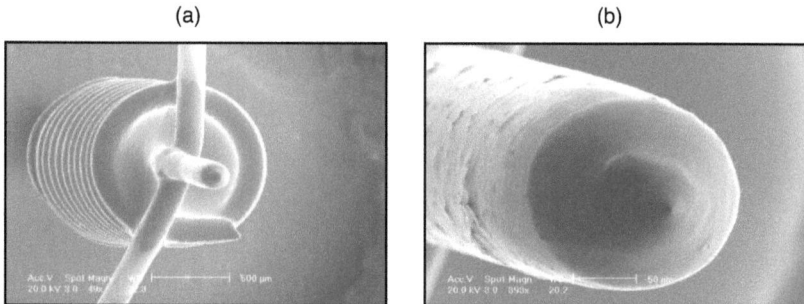

Figure 5.3: Magnification of the liquid metal ion source used in the FIB system. In reality the size of the ion source is about 7 mm. The Ga reservoir of the specific model used for this work lasts for about 1000 hours of operation.

fields. The LMIS shown in figure 5.3 (a) consists of a coil as Ga reservoir and a sharp pin made out of tungsten. The very sharp tip at the end of the pin is dynamically created during operation due to back-sputtering of tungsten. In 5.3 (b) a magnified image of the tungsten pin is shown. The tungsten pin is covered with a thin layer of Ga from the reservoir. The high electric fields shape the Ga layer of the very end of the pin to a so called *Taylor cone* [93], where the Ga^+ ions will be extracted finally. The design and the effects that lead to the formation of the tailor cone guarantee the extraction of Ga^+ ions from a very small spot size, which is essential for accomplishing a focussed beam diameter in the order of 10 nm.

Although the ion column containing the ion optics is optimised for an acceleration voltage of 50 kV, it is possible to vary the acceleration voltage between 5 kV and 50 kV. The advantage of using lower ion energies than 50 kV is that the ions will not penetrate too far into the sample, which is important for sensitive samples. However, with decreasing ion energy the ability to focus the ion beam decreases also. Therefore the achievable resolution of the local sputter process decreases.

The *extractor* extracts a broad Ga^+ ion beam (spray) from the source (figure 5.2). Most of the ions will be absorbed by the *spray aperture*, which is used to sense the ion beam intensity. The *suppressor* counteracts the ion extraction and is used to adjust the ion intensity via an active feedback from the spray aperture. This is of high importance, because the ion emission

---

allows application in UHV due to negligible Ga gas pressure at (near) ambient temperature conditions.

from the source changes during operation. The small fraction of ions which are not absorbed by the spray aperture will be focussed by *lens 1* and *lens 2*. Lens 1 is used to focus the ion beam on the *beam limiting aperture*, which is responsible for further reducing the beam diameter and intensity. One can choose between different aperture diameters. The smaller the aperture diameter, the smaller the final beam diameter and the smaller the beam intensity. Note that the beam limiting aperture has to be replaced from time to time because the aperture diameters widen during operation due to sputter processes induced by the ion beam.

To reduce the effect of mutual repulsion of the Ga^+ ions, the system was constructed in a way that single ion trajectories never intersect. This guarantees a small beam diameter. The *stigmator unit* consists of eight deflection electrodes and is used to correct aberrations from the ideal symmetric beam shape.

After the beam has passed the beam limiting aperture, it reaches the *beam blanker*. The beam blanker is used to deflect the ion beam to the *beam blanking aperture*, where the final beam intensity is measured. Unless the beam is blanked, it reaches the *deflection electrodes* which are used to move the ion beam over the sample in a desired way.

The *multiple channel plate* is located at the bottom of the ion column and is used to detect secondary electrons or ions which are emitted by the sample spot hit by the ion beam. In this manner, images of the sample can be recorded similar to scanning electron microscopes.

FIB systems are not limited to simply sputtering away material. The range of application can be further extended by using different gases that adsorb to the sample surface. It is possible to conduct selective, local etching processes similar to reactive ion etching (RIE). Another possibility is to *deposit* material like in chemical vapour deposition (CVD) processes (see chapter 4.2.3 for further information). However, in contrast to conventional CVD, the adsorbed precursor decays only on locations where the ion beam hits the surface, enabling very local material deposition with a resolution near the focussed ion beam diameter.

5.3 Sample Preparation and Experimental Considerations

As samples p-doped silicon wafers with an acceptor concentration of $N_A=9.4\times 10^{14}$ cm^{-3} were used. During FIB-processing, Ga^+ ions (acceptors) were implanted into the sample. Therefore, p-type silicon was chosen to avoid any additional difficulties in data interpretation due to the very complex electrical behaviour of pn-junctions[29, 86, 87], which would emerge when n-type semiconductors were used. The rather low acceptor concentration of the bulk material was an advantage, because SCM yields higher signals on low doped semiconductors (refer to figure 2.7). The samples were prepared in two ways to match the different demands of FIB shape and damage depth determination:

The very clean surfaces of freshly cleaved silicon wafers were used for the beam shape determination in chapter 5.4. On the cleaved surface five types of spots were made with the FIB, which differed from each other in the deposited ion dose (0.025, 0.05, 0.1, 0.5 and 5 pC/spot). The spots were located in close vicinity (a few microns) of the wafer edge in order to reduce sample tip-holder overlaps and to minimize the related impact of stray capacitance on the SCM measurements3.1. The acceleration voltage of the Ga^+ ions was 50 kV. The aperture size of the FIB system was 50 μm with a constant ion current of 50 pA.

For the investigation of the damage-depths below FIB irradiated areas in chapter 5.5, a trench was milled into the polished front side of a wafer. The length of the trench was 1 cm and the depth was 2 to 3 μm to be easily visible in the optical microscope. Note that the biggest aperture size in the FIB machine of about 400 μm with a related ion current of 25 nA had to be used to hold process times tolerable when milling such a long and deep trench. To investigate the damaged area below the trench, the sample was cleaved. An illustration of the cleaved, FIB processed sample is given in figure 5.4.

So, the SCM measurements for both FIB shape as well as for damage-depth determination were carried out on the cleaved silicon surfaces. The AFM-probes used for this investigations were conductive diamond tips.

Before the results are presented, the contrast mechanism in the SCM images has to be discussed briefly. As shown in chapter 2 (also refer to [29, 86, 87]), SCM contrast is a complex function of the semiconductor's doping concentration and the applied tip-bias voltage. On unipolar doped silicon (in the absence of pn-junctions), however, the tip-bias voltage can be adjusted in a way that the SCM signal *decreases* monotonically (roughly exponentially) with *increasing* doping level (refer to chapter 3.5 or the original article [105]). However, besides the p-type doping by Ga^+ ion implantation, the crystalline structure of FIB irradiated samples is damaged heavily. In general, any damage will yield a reduction of the SCM signal and therefore, as a first step, no effort was made to distinguish between doped, amorphisized or damaged areas. On the other hand it should be emphasized that ion beam

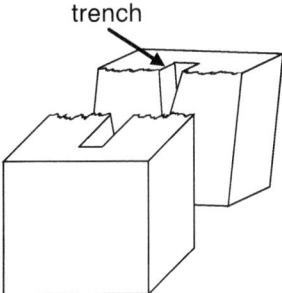

Figure 5.4: Illustration of a FIB made trench on a sample which is cleaved into two pieces.

doses which are magnitudes smaller than the minimum dose for surface modification (e.g. swelling due to a smaller material density inside amorphisized regions) can already be detected via SCM, which is shown in the following chapters. Besides, studies on this topic have also been performed in literature with lithographically patterned silicon samples after implantation of arsenic (As) with an ion implanter [114].

5.4 Ion Beam Profile – Damage on Si-Surfaces

To determine the beam radius of the FIB system via SCM, single implantation spots were used. Figure 5.5 (a) shows the resulting topographic changes of milling with a moderate dose per spot (0.5 pC/spot). The swelling and subsequent silicon removal due to sputtering was reproduced very well and leads to the typical crater-like structures. The inset in figure 5.5 (a) shows a magnification of the relevant area. Applying lower ion doses leads to hillock shaped structures because swelling dominates sputtering. Figure 5.5 (b) shows the simultaneously measured SCM picture. As one can see, very good contrast between the FIB-modified and other sample areas was obtained. In comparison with the topographic image seen in figure 5.5 (a), the simultaneously recorded SCM picture in (b) shows a significantly larger damaged region. The impact of the ion irradiation on the silicon sample can now be used to measure the ion beam radius. However, this leads to two possible definitions of a beam radius since there are two different responses to ion irradiation:

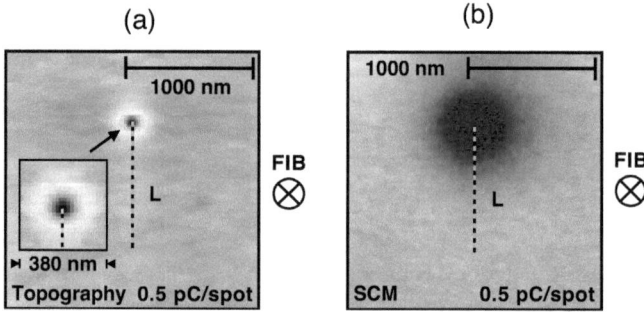

Figure 5.5: (a) The topographic image of a FIB irradiated spot shows a crater like structure. The insert shows a magnification. The beam direction is indicated. (b) Corresponding SCM image.

- First, one can look at ion induced changes in topography (figure 5.5 (a)). Depending on the ion dose, the irradiated spots can look like craters or hillocks. Therefore, the point where the outer swelling flank (craters also have inner flanks, this inner flanks are not of interest here) has decreased to 50 % was taken as the outside border of the FIB induced structure (this is shown in figure 5.6 (a)). The corresponding radius of the topographic structure and subsequently the beam radius is named R_{Topo}.

- Second, one can take the SCM-signal to define the beam radius R_{SCM}, which is half the distance between the points where the SCM signal flanks rise to half of their maximum (figure 5.6 (a)).

The two radii R_{Topo} and R_{SCM} are compared in figure 5.6 (a), where cross sections of the FIB irradiated areas are plotted along line L of figure 5.5. The difference between topography signal and SCM signal, $\Delta R = R_{SCM} - R_{Topo}$, is 253 nm. Figure 5.6 (b) compares the behaviour of the radii R_{Topo} and R_{SCM} with increasing dose per spot. The radii show monotonic growth, however, there are saturation effects in the high dose regime. In addition, both data sets diverge for big ion doses. Whereas for the lowest dose the ion irradiated spots have twice the radius in SCM mode than in topography mode, for the highest dose this ratio is almost four. Our observation, that the SCM based beam radius R_{SCM} is always larger than the topographic radius R_{Topo}, and the effect that R_{Topo} and R_{SCM} diverge for big ion doses can be explained by the following facts:

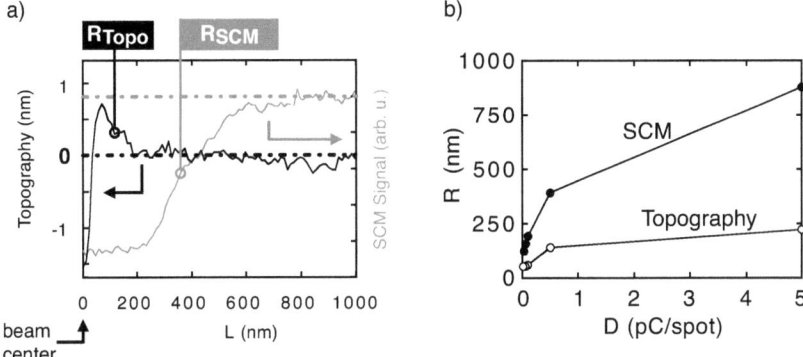

Figure 5.6: (a) Radial cross sections through the topographic and SCM image along the line L. R_{Topo} and R_{SCM} define the radius of damage as seen in the topographic and the SCM image. (b) R_{Topo} and R_{SCM} versus ion dose per irradiated spot.

First, as was already published [75, 7, 96] the beam profile consists of (at least) two regions: The region far away from the beam centre, where the overall intensity is very small but decays slowly. The other region is close to the beam centre, where the intensity is comparable with the beam centre and has the steepest decay. In figure 5.7 a scheme of the expected beam radius is shown.

Second, SCM is much more sensitive to ion irradiation effects than the topographic signal, since topographic changes by amorphisation need very high ion doses. In other words, the threshold for amorphisation is higher than the threshold for an impact on the SCM signal. Therefore, as can be seen in figure 5.7, topographic changes rather map the centre of the beam, whereas changes in the SCM signal map the beam periphery.

Because of these two properties the SCM sensed beam radius R_{SCM} grows quickly with increasing dose per spot in the outer areas of the beam profile, whereas the smaller crater-like structure in the topography grows just slowly.

5.5 Ion Beam Damage Inside Si-Samples

A second important subject for FIB application is, as already mentioned in the chapter 5.1, the determination of the width and depth of FIB induced damage *below* the sample surface. As explained in chapter 5.3, a long trench

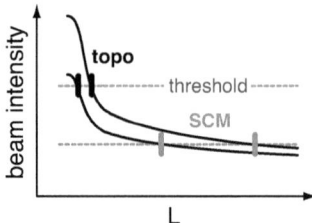

Figure 5.7: Focussed ion beam profile for two beam intensities. The beam centre is located on the left side. An increasing beam intensity leads to different growth rates of the observed effects, depending on the detection limits.

was milled into a Si sample. Afterwards the sample was cleaved to expose the ion damage round the trench inside the sample.

Figure 5.8 (a) shows the topography of this cleaved, FIB made trench. The corresponding SCM signal is shown in figure 5.8 (b). Again, the tip voltage was optimized to obtain the largest SCM signals in non-irradiated, low doped areas. The trench's impact on the SCM signal exceeds the actual dimensions of the trench. With a certain probability, ions can be scattered out of their incident direction and move perpendicular to the beam. In that way, they can reach areas not covered by the beam area.

Figure 5.9 shows a scheme of the FIB made trench. The grey shaded area corresponds to ion beam damage, which is located in the vicinity of the trench. For the subject of device modification it is important to know the size of the damaged region below and at the edge the trench, corresponding to the values A_{FIB} and B_{FIB} of figure 5.9.

An estimate of A_{FIB} and B_{FIB} can be obtained by comparison of the topographic and SCM data of a trench. Figure 5.10 (a) compares the topographic and the SCM signal height plotted along line $L_\parallel$ (figure 5.8) parallel to the incident beam. The distance A_{FIB} between the topography signal and the SCM signal is approximately 620 nm. Figure 5.10 (b) shows plots of the topographic and the SCM signal height along line $L_\perp$ perpendicular to the FIB direction. The distance B_{FIB} between topography signal and SCM signal is about 310 nm. A comparison of figure 5.10 (a) and (b) shows a ratio B_{FIB}/A_{FIB} of approximately 1/2. This ratio is also confirmed by transmission electron microscopy (TEM) investigations of FIB irradiated MOSFETs [4]. Figure 5.11 (a) show a TEM image of the polysilicon-gate of a MOSFET after FIB treatment[2]. To explain what can be seen in figure 5.11 (a), the

---

[2] A trench was cut into the passivating layer and the polysilicon gate material to allow

CHAPTER 5. FIB INDUCED DAMAGE INVESTIGATED BY SCM   91

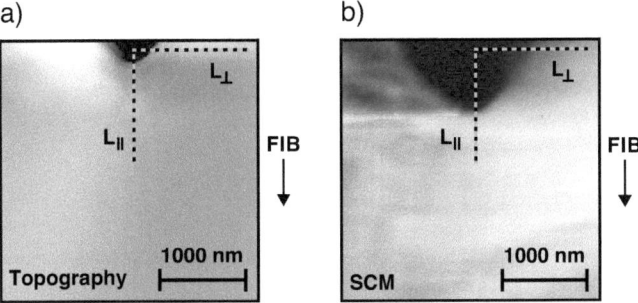

Figure 5.8: (a) Topographic image of a FIB milled trench in cross sectional view. The beam direction is indicated. (b) Simultaneously recorded SCM image. The lines $L_\parallel$ and $L_\perp$ indicate the directions where the topographic as well as SCM signals will be compared to calculate the extent of the ion damage.

special FIB milling process has to be taken into account. In the case of figure 5.11 (a), the ion beam was scanned many times over an area of 10×0.2 μm^2 until the desired depth of about 2 – 3 μm was reached. The depth is a function of the deposited ion dose. During a single scan over the chosen area, only a small amount of the final ion dose was deposited, and only a small amount of the surface was sputtered away[3]. For clarity, a scheme of the scan strategy is shown in figure 5.11 (b). However, a certain amount of the sputtered material will be redeposited again on the sample near the point it was sputtered away. Furthermore, the deeper a trench is milled, the more material will be redeposited on the side walls of the trench. Therefore, the diameter of the trench at the bottom is only 30 nm (at the top it is still 0.2 μm) and there will always be a layer of redeposited material. In figure 5.11 (a), one can see two different layers of material at the bottom and the side walls of the trench. The side wall of the trench consist of redeposited material, followed by a halo of strongly damaged or amorphisized material.

Again, one can determine the extent of the ion beam damage. In contrast to the SCM investigations of cleaved silicon wafers, the TEM image of the MOSFET shows a damaged area that extends only about A_{FIB}=200 nm in the beam direction and about B_{FIB}=100 nm perpendicular to it. This is

---

local dopant implantation via an implanter. Ions can only be implanted into the MOSFET channel at the lowest regions of the trench, whereas elsewhere on the chip they are masked by the passivating layer.

[3]In another possible scan mode, the whole ion dose is deposited in one single pass. The different scan modes have a tremendous impact on the final shape of a trench.

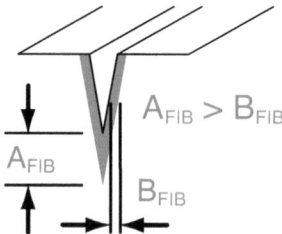

Figure 5.9: Scheme of a FIB milled trench. The grey shaded area corresponds to the ion damage done to the sample. Because of the low probability that an ion is scattered off the incident beam a large angle, the damage at the side walls (B_{FIB}) extends less into the sample than the damage at the bottom (A_{FIB}) of the trench.

only one third of the damage extent measured by SCM on the cleaved Si bulk samples, even if the ratio B_{FIB}/A_{FIB} is also 1/2. This can be explained by comparing the detection sensitivity of TEM and SCM. To get contrast in TEM, a crystalline substrate has to be amorphisized to a high extent, which needs very high ion doses. On the other hand, SCM is able to detect impurity concentrations down to 10 – 100 atoms per μm^3 (10^{13} - 10^{14} atoms per cm^3), which is magnitudes more sensitive than TEM.

CHAPTER 5. FIB INDUCED DAMAGE INVESTIGATED BY SCM 93

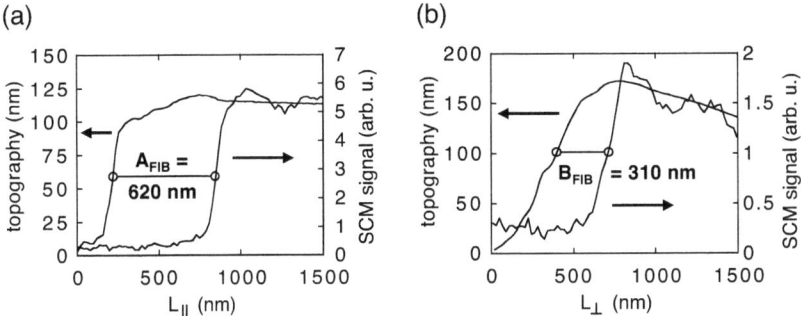

Figure 5.10: (a) Comparison of topographic and SCM signal height along the line $L_\parallel$ parallel to the ion beam. (b) Topographic and SCM signal height are also compared along the line $L_\perp$ perpendicular to the incident beam. In this direction, the difference between topography and SCM is significantly less than in (a).

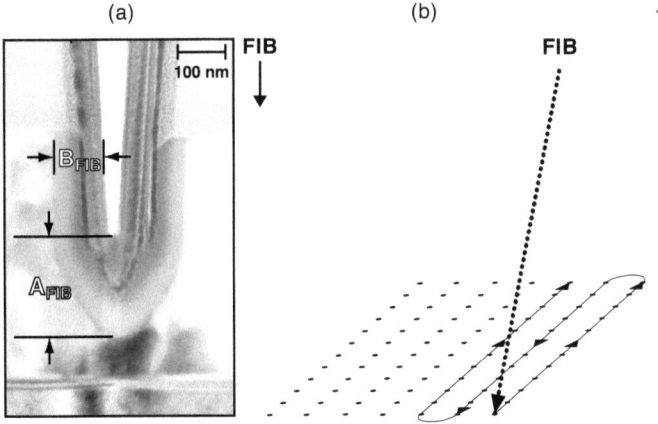

Figure 5.11: (a) TEM image of a trench milled into the polysilicon gate of a MOS-FET by a FIB process. One can easily distinguish between redeposited material inside the trench and the amorphisized halo further away from the trench. A_{FIB} and B_{FIB} are the sum of the redeposited material and the amorphisized areas in the corresponding directions. (b) Milling scheme of the FIB system.

Chapter 6

Microscopic Capacitance Spectroscopy of SiO$_2$ and ZrO$_2$ Covered Samples

6.1 Introduction

All data presented in this chapter were measured by the QSCS-setup described in chapter 3.2. The main focus in chapter 3 was to describe the QSCS-setup and to show the principle of operation of a capacitance bridge and the difference to conventional SCM. This part first develops a useful set of parameters and experience to run the setup effectively. Afterwards, to show the very good quality of the nanoscopic C(V) curves obtained by QSCS a comparison of nanoscopic C(V) curves and C(V) curves from large scale MOS capacitors was performed. Finally, to show that QSCS-setup can be applied to get useful information about a sample, the method was used on ZrO$_2$ covered samples that show very small scale growth variations.

6.2 Developing a Useful Parameter Set

6.2.1 Pointwise Averaging – Average Time Level

An important parameter of the capacitance bridge, which can be accessed by the user, is the *average time* level. It determines how many single capacitance measurements are averaged internally to obtain the resulting value which is then displayed on the bridge itself or is transmitted to a computer. Recording more capacitance values means longer measurement duration, hence the name *average time*. For the sake of completeness and as a reference, the

CHAPTER 6. MICROSCOPIC CAPACITANCE SPECTROSCOPY

time consumption[1] at a certain average time level is shown in table 6.1 taken from the bridge's manual[1]. The duration approximately doubles with every increment of the average time level. The larger the average time level, the less noise is superimposed to the final capacitance signal. If the noise is truly random, the increase of measurement accuracy is proportional to the square root of the measurement duration, which is a well known relation from statistics. Figure 6.1 shows five C(V) curves recorded with average time levels ranging from 8 to 12 with other parameters held constant. One can see the improvement of the signal to noise ratio with increased measurement time. To compare the noise of the C(V) curves, the standard deviation σ of the

Table 6.1: The average time level as it is fed into the capacitance bridge and the corresponding measurement duration in seconds.

average time level	approximate measurement time (s)
8	2.5
9	4.0
10	6.9
11	12.0
12	23.0
13	44.0
14	91.0
15	188.0

grey shaded region on the left half of the curve was determined. The reasons which led to the decision to take an average duration of 6.9 s per measurement point was a simple tradeoff between accuracy and time consumption. Especially during the very first experiments the lateral and vertical drift of the AFM tip posed some problem because it distorts the C(V) curves. The development of methods to reduce the AFM tip drift is described in chapter 6.3. The longer a singe C(V) curve takes to be recorded at a constant drift rate, the more the curve will be distorted. All of the following C(V) curves presented here were therefore recorded taking an average duration of 6.9 s per measurement point, regardless of later improvements to reduce tip drifts.

---

[1]The capacitance bridge uses some 'intelligent' software routines to determine if the desired measurement accuracy can be reached. In case the noise level is too high or the capacitance of the sample is changing quickly, the bridge starts the measurement again hoping for the conditions to improve. Therefore, in general, the measurement duration for a single measurement is unknown. The *approximate measurement time* given in table 6.1 is the duration of a single measurement in the case the first attempt is successful.

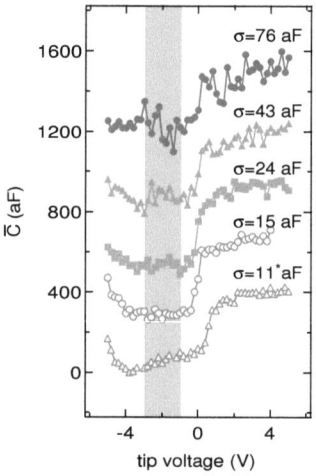

Figure 6.1: The influence of the average time level on the accuracy of the recorded C(V) curves. From top to bottom: average time level 9, 10 11 and 12 corresponding to a time consumption per point of 2.5s, 4.0s, 6.9s, 12s and 23s, respectively. To separate the curves, offsets were added. The standard deviation σ was calculated for every curve in the grey shaded area. To calculate σ for the curve at the bottom of the graph (time consumption: 23s), the parasitic constant slope at the left half of the curve was subtracted. Measurements by courtesy of Matthias Schramböck.

Furthermore, a delay of 7 seconds were included in between the single points for equilibration purposes. Therefore, one capacitance value every 14 seconds is put out by the bridge. Where it seemed necessary, multiple C(V) curves were recorded on one and the same sample spot. Afterwards all this curves were averaged pointwise to increase the accuracy (see chapter 6.4 for further information).

6.2.2 The Modulation Voltage

Another important parameter is the amplitude of the test voltage (*modulation* or *excitation voltage*) applied by the bridge during a measurement. The bridge allows to input an applied *maximum* of the excitation voltage in the range of 0.3 mV to 15V [1]. However, the actually applied excitation voltage may be lower than the user input depending on the magnitude of the measured capacitance. In general, larger capacitance values lead to excitation voltages which are lower than the maximum value given by the user. Table

Table 6.2: Excitation voltages automatically applied by the bridge when connected to a capacitance of the given magnitude. However, user input overrides every attempt of the bridge to apply larger voltages than the user value.

Limit (volts)	Maximum capacitance (pF)
15.00	110
7.50	220
3	550
0.25	6600

6.2 taken from the bridge's manual [1] summarises that rule. For example, a user input of 15V and a tested capacitor of more than 6600pF would lead to an excitation voltage of only 0.25V. However, the user input of a certain excitation voltage value overrides every attempt of the bridge to apply larger voltages than the user input in case of a small capacitor. Therefore it is possible to measure a 1pF capacitor with an excitation voltage of, say, 0.1V. To understand how the applied excitation voltage is calculated is crucial for several reasons. First, some samples used here need a limitation of the totally applied voltage because of their low brake-down voltages of less than 5V. Another reason for a reduction is that higher excitation voltages change the shape of the C(V) curves. With increased excitation voltage the C(V) curves become broader. The reason for this behaviour can be explained as follows: If the applied constant DC voltage operates the MOS capacitor near the flatband voltage but is still in accumulation, a large AC excitation voltage, during parts of the cycle, will drive the capacitor beyond the flatband condition into the depletion regime. The net-effect is a capacitance value somewhere between accumulation and depletion regime. Figure 6.2 shows the influence of different excitation voltages on the quality and shape of the C(V) curves. There is no visible broadening of the curves when the excitation voltage is increased from 0.1V to 0.25V. However, in figure 6.2 one can also see a decrease of the noise when the excitation is increased. Higher excitation voltage means more oscillating charge in relation to random charge fluctuation, and therefore an increased signal to noise ratio. At an excitation voltage of 3V the curve appears very smooth but is much too broad to extract any useful data. It is a tradeoff between noise and broadening. Because lower excitation voltages than 0.25V do not increase the slope of the transition region between accumulation and inversion in figure 6.2 and only lead to more noise, it made sense to use an excitation voltage of 0.25V throughout the following experiments.

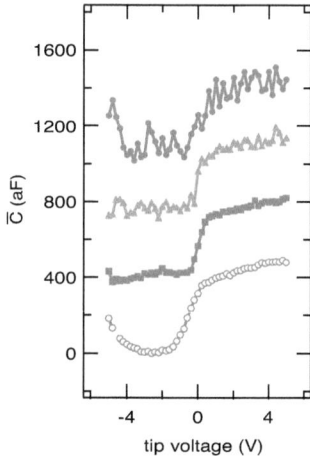

Figure 6.2: The influence of the modulation voltage on the accuracy and the shape of the C(V) curves. From top to bottom: modulation voltages of 0.1V, 0.25V, 0.75V and 1.5V . To separate the curves, offsets were added. In general, higher modulation voltages lead to a reduced noise, but broaden the C(V) curves. Measurements by courtesy of Matthias Schramböck.

6.2.3 Does QSCS Influence the Sample?

The next important question is whether the QSCS measurements do influence the sample surface or not. As described in chapter 4, charging of the oxide covered sample surface is a severe problem for standard SCM methods. Under adverse conditions charging can destroy the SCM signal within minutes when scanning over an area. Even if dielectric breakdown can be avoided utterly due to the exact determination of the bridges' electrical output voltage, there remains the possibility of electrochemical processes which could degrade the oxide cover of the sample. Especially the fact that in QSCS the AFM tip often remains a very long time on one and the same sample spot makes it possible that even a very small eventual degradation rate can accumulate to a severe effect. To investigate that issue, the approach was similar as in chapter 4, where a visible degradation of the oxide was produced by scanning over one and the same area several times. For the similar QSCS experiment, about 120 single C(V) curves were recorded subsequently on one sample spot of a SiO_2 covered Si sample. The thickness d of the oxide was only 4 nm. Afterwards, packages of 30 single C(V) curves were averaged pointwise to

CHAPTER 6. MICROSCOPIC CAPACITANCE SPECTROSCOPY

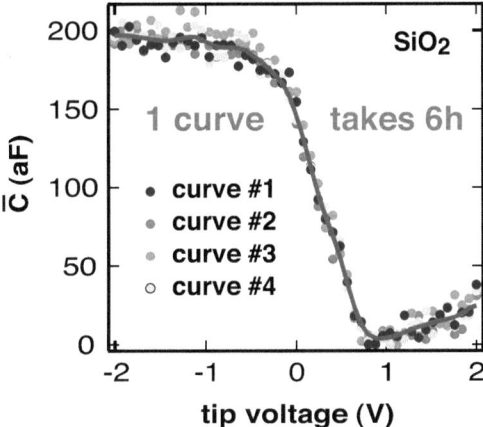

Figure 6.3: 4 C(V) curves obtained on one and the same sample spot on SiO_2 after averaging over 30 single curves. There are no differences between the curves, therefore one can assume that within 24 hours of continuous QSCS-measurements the sample surface does not degrade.

obtain 4 resulting curves, which are shown in figure 6.3. It took about 6 hours to obtain the 30 single curves for one of the averaged curves in figure 6.3. Neither a change of the oxide (accumulation) capacitance nor a shift of the transition region (flatband voltage) indicating degradation of the oxide or some parasitic build-up of charge inside the oxide. All the 4 averaged curves in figure 6.3 have the same shape and position. This is the verification that QSCS measurements under moderate voltage conditions do not influence the properties of the sample.

6.3 AFM Tip Drifts and Countermeasures

At the current state of development the QSCS-setup runs without any laser feedback, to get rid of the influence of light. Of course, there are some possibilities to implement feedback loops that do not influence the C(V) measurements. Approaches that supply control for the up and down movement of the tip (z-direction) are for example AFM sensors that detect the tip deflection via tunnelling currents (like the first AFM back in 1986 [13]), or the usage of laser wavelengths that do not interfere with the sample (e.g. photon energies lower than the band gap of the semiconductor sample). However, all of that approaches are more or less difficult to implement, because they require major reconstruction of the AFM hardware. Therefore the decision was to simple switch off the laser feedback to get rid of the parasitic light induced effects inside the sample. Without any compensating feedback, AFM tip drifts became an issue. Especially drifts in z-direction are annoying. Either the tip moves up (away from the samples surface) and finally the contact to the sample is lost, or it moves down and may damage the dielectric layer. If QSCS may be applied for Schottky capacitance measurements, even much smaller z-position changes than generally observed will lead to a change of the capacitance behaviour of the tip sample system due to an anticipated pressure dependence of the schottky contact[17].

This chapter will discuss the different causes of drifts, and how they can be prevented or compensated. There are 2 reasons for tip drifts which could be observed in this work.

6.3.1 Piezo Creep

First, every piezo suffers from so called *piezo creep* [52]. If a voltage change is applied to a piezo, the piezo will expand or retract corresponding to the voltage. However, instead of reaching its new length (or position) suddenly, it asymptotically moves to the new equilibrium position. This occurs both in lateral (x- and y-) as well as in vertical (z-) direction. The magnitude of the creep is a few percent of the piezo expansion. The observed time constant is a few minutes for the Digital Instruments AFM used in this work. A lot of efforts have been made to get rid of [39], or compensate for the non-ideal behaviour of piezos [46, 5, 113]. The easiest possibility to get rid of this type of tip drift is simply to wait long enough until the piezo has nearly reached its equilibrium position prior to any measurements. This was the approach taken for this work. The larger the initial position change (respectively voltage change to the piezo) the more and the longer the piezo creep occurs. Therefore one useful measure is to avoid any large piezo movements which reduces piezo creep time. Another possibility to shorten creep time is to apply a slightly larger voltage than required to reach a certain position. The piezo will expand a little more and would eventually overshoot the desired position. After a few seconds the voltage is again reduced. If this procedure is done properly, the piezo does not show creep anymore. The reason for that behaviour is that creep is a *slow* change of internal polarisation of the piezo crystal. Driving the piezo at slightly higher voltage for a short time means to apply a degree of polarisation to the piezo which is otherwise reached only after a considerable longer time due to piezo creep.

CHAPTER 6. MICROSCOPIC CAPACITANCE SPECTROSCOPY   101

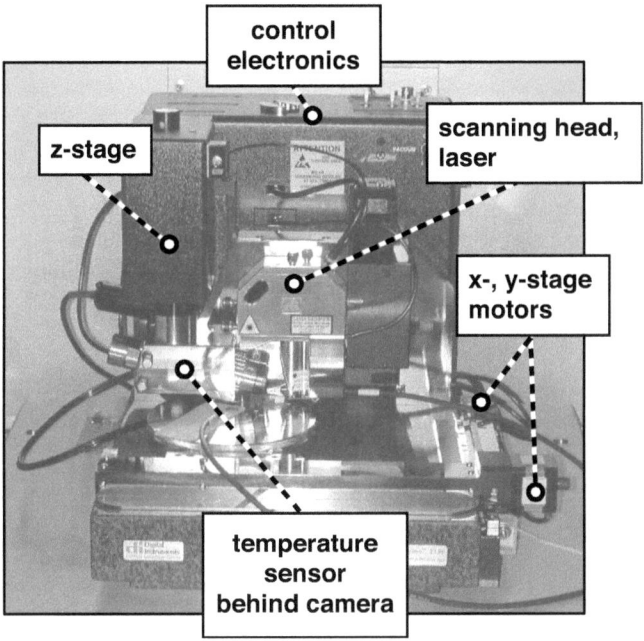

Figure 6.4: Image of the AFM inside the acoustic hood. Both the positions of all relevant heat sources as well as the location of the temperature sensor is marked.

6.3.2 Thermal Expansion

The second, much more severe reason for tip drifts is due to thermal expansion of the setup. As mentioned in chapter 3.2.2, the AFM and the sample must be kept in complete darkness inside an acoustic hood. Due to heat dissipation inside the AFM electronics, the temperature inside the hood is about 15 °C higher than the ambient temperature when the hood was closed for a long time (a few days). The setup for temperature monitoring was already described in chapter 3.2.2. It is important here to understand the relative complex interaction between the (location of the) temperature sensor and the different heat sources of the AFM hardware inside the hood. There are different locations inside the AFM where electricity is dissipated to heat. Figure 6.4 shows a scheme of the AFM hardware inside the acoustic hood together with the positions of the temperature sensor and all heat sources. The sensor only measures the *air temperature* at a single spot inside the hood. One can presume this single spot to be thermally shielded to a certain amount from the other parts of the AFM. The main heat sources which could be identified are: the control-electric at the backside of the AFM, the video camera near the temperature sensor, the motors of the x-, y-, and z-stage and the laser inside the scanning head. Whereas the control electronics and the camera are heating the setup continuously, the other heat sources are only temporarily switched on and off (this causes more trouble, which is explained later). Because the different heat sources lead to different thermal expansion of different parts of the AFM, a net increase of the air temperature measured by the sensor at a certain location can lead to both an increase as well as a decrease of the tip-sample distance, depending on which heat sources are switched on. The author is aware of the fact that the presented knowledge on this important subject is very fragmented. There is much room for improvement, many aspects of thermal expansion have to be measured in future measurements. However, having at least a semi-quantitative idea of the movement of the scanning head (e.g. the prediction if the head drifts up or down) gives a big advantage in trouble shooting (e.g. no signal because tip drifted away from sample) and opens the possibility to compensate for the drift.

Lets start with the simplest case and assume that only the camera, the control electronic and the laser dissipate heat, and the dissipation rate does not change with time (e.g. laser is *not* switched off after a while) Figure 6.5 (a) shows the temperature change during 30 minutes inside the (closed) hood after the hood was open for a long time (about 20 minutes). The grey shaded area marks the last 10 minutes in the plot where the temperature change is about 9mK/minute. Figure 6.5 (b) traces the temperature change for an extended time interval of about 16 hours. After twelve and a half hours (750 minutes) the temperature change is just about 0.29 mK/minute. But how can this temperature changes be interpreted in terms of nm AFM tip drift? Unfortunately, it is only possible to present field reports due to a lack of systematic studies. As a rule of thumb it turned out during the experiments, that a *slow* temperature increase of 1 mK leads to *raising* the AFM tip about 1 nm. This would lead to a vertical tip displacement of about 108nm (!) while recording one single C(V) curve in about 12 minutes. Such a large displacement makes it virtually impossible to perform any useful measurements! However, after thermal equilibration for about 13 hours (see figure 6.5 (b)), the displacement is only 3.5 nm for a single C(V) curve. However, for a sufficiently stable temperature one has to wait at least one day. Especially when recording large numbers of curves on one and same sample spot for averaging purposes, it is recommended to wait 3 days. It is important to note that even opening the hood for only a short time (about

CHAPTER 6. MICROSCOPIC CAPACITANCE SPECTROSCOPY

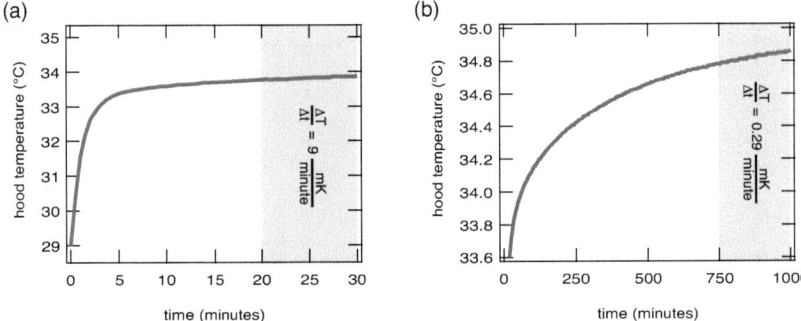

Figure 6.5: Temperature change T_{sensor} inside the acoustic hood when closed a) for 30 minutes and b) for 16 hours measured with the temperature sensor behind the camera system. For both plots the rate of temperature change was calculated from data inside the grey shaded area.

1 second) introduces a sudden temperature decrease of a few Kelvin (!), and it takes at least one hour for the temperature to reach the equilibrium again. The next aspect that has to be explained is why the AFM tip moves *up* when the temperature is increased. One can exclude the possibility that parts of the AFM hardware have a negative linear thermal expansion coefficient. An explanation why the tip is raised when the temperature is increased is that the scanning head still pushes the AFM tip down due to thermal expansion, but the larger z-stage also expands and lifts the whole scanning head together with the AFM tip. It depends on the relative size of the z-stage and the scanning head whether the net effect on the tip is a moving up or a moving down. The vertical displacement Δl of the tip can be calculated via the following equation

$$\Delta l = l_{head} \times a_{head} \times \Delta T_{head} - l_{stage} \times a_{stage} \times \Delta T_{stage}, \quad (6.1)$$

where a negative value for Δl means the tip moves up and a positive value means the tip approaches the sample. Assuming the same temperature change $\Delta T = \Delta T_{head} = \Delta T_{stage}$ for all parts of the AFM (isothermal case) and the same expansion coefficient $a = a_{head} = a_{stage}$, still the size of the z-stage is larger than that of the scanning head and therefore the tip moves up with increasing temperature. Taking the expansion coefficient of steel $a \approx 14 \times 10^{-6}$ (average from different internet sources) and assuming the size of the z-stage and the head to be $l_{stage} = 20$ cm and $l_{head} = 10$ cm, a temperature increase of $\Delta T = 1$ mK leads to a tip displacement of $\Delta l = -1$ nm.

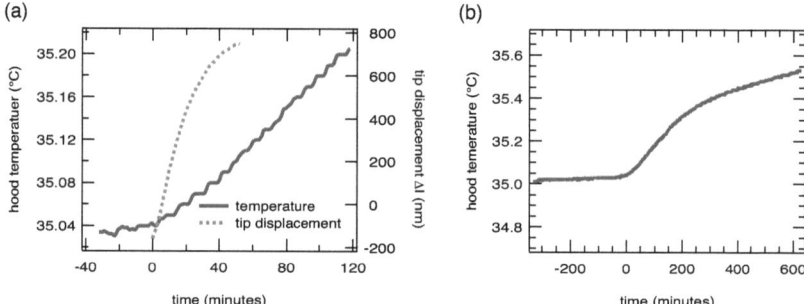

Figure 6.6: a) tip displacement and temperature increase after switching the laser on. b) temperature increase at a larger timescale. Zero at the time axis correspond to the moment of switching the laser on. The hood temperature (T_{sensor}) was again measured with the temperature sensor behind the camera system.

Observations during the measurements show approximately the same value. The influence of such vertical movements on the actual C(V) measurements depends strongly on the geometry and spring constant of the used cantilever, which will be discussed later.

As discussed in chapter 3.2.2 (footnote on page 47), all C(V) measurements have to be done in complete darkness. Therefore switching off the feedback laser is mandatory. This means that the scanning head (where the laser is located) is no longer heated, and the temperature T_{head} of the scanning head eventually decreases more quickly than the temperature T_{stage} of the z-stage. Due to equation 6.1, this means that the tip moves up. One easy way to record thermal expansion or contraction is to simply record the counteracting movement of the z-piezo which is controlled by the AFM laser feedback system (which means that the laser has to be switched on). In figure 6.6 both the time evolution of the thermal tip displacement Δl as well as the air temperature was recorded when the laser was switched *on* after the system got used to heating conditions when the laser was switched *off* for a long time. The thermal tip displacement Δl was measured by recording the piezo voltage applied by the feedback system to counteract thermal expansion of the scanning head. The displacement Δl is positive, indicating that the tip moves *down* due to a suddenly increase of the temperature T_{head} inside the scanning head. The initial tip displacement rate is about +30nm/minute. At the last data point the recorded displacement is about $\Delta l = +840$ nm, whereas a calculation based on the data of the temperature sensor assuming isothermal heating (temperature changes occur everywhere at the same time and the same rate) gives a displacement of Δl = -98nm. Whereas the tip displacement nearly reaches its equilibrium after 50 minutes, the air temperature still increases constantly. In figure 6.6 (b) one can see that the air temperature at the sensor's position nearly reaches its equilibrium only after 10 hours. If the presented expansion model is valid, it is possible that the tip displacement could reverse sign when the temperature raises also sufficiently in the z-stage. This is be a topic for future investigations.

Another important aspect is that every vertical movement also creates a lateral movement if the tip is in contact with the sample. This is a simple geometric effect, because the AFM cantilever is attached to the tip holder at an angle of 13° with respect to a horizontal line. Figure 6.7 shows what happens if the AFM tip is in contact with the sample and the piezo is expanded further to increase the pressure between the tip and the sample. In Figure 6.7 (a) the tip has just made contact with the sample, and the angle between the cantilever and the sample is still γ = 13°. In figure 6.7 (b) a simple model is presented which describes the result of increasing the tip-sample pressure by expanding the z-piezo (either electrically or by thermal drift). γ is reduced, and therefore the projection of the tip on the sample $cos(\gamma)$ is increased. Simple trigonometric calculations show that even a relatively small expansion of the piezo of only 50nm leads to a lateral displacement of 11 nm (for the 225 μm long cantilevers which were used for this work). However, this is an oversimplification, because in reality, the cantilever bends under pressure, which is shown in figure 6.7 (c). Therefore, if high spacial resolution is desired, one has to compensate for thermal drifts. Fortunately, larger lateral drifts can be observed during C(V) measurements as changing stray capacitance values $\overline{C}_{stray}$. Especially when performing C(V) measurements near the edge of a sample lateral movements of the AFM tip effectively changes the surface of the parasitic stray capacitor. An increase of the stray capacitance because of a decrease of the distance between the electrodes of the stray capacitor is negligible. The distance between the sample and the cantilever bulk is about 50 μm whereas distance changes due to drift is typically about 50 nm. This scarcely influences the stray capacitance. In figure 6.7 (d) C(V) curves were recorded on successive locations about 1 μm apart. The curves recorded near the sample edge typically showed smaller stray capacitance than the curves recorded at locations far away from the edge. In this particular experiment, the capacitance decreases about 100 aF when the tip moves 1 μm nearer to the sample edge. In general, this quantitative correlation will not hold for other experiments. However, one can use this correlation to test for unwanted lateral drifts. Whenever a changing stray capacitance indicates a lateral tip movement, the measured data should be

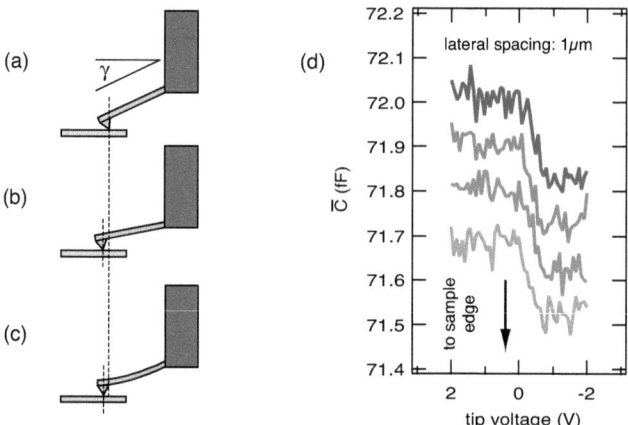

Figure 6.7: Influence of the tip deflection on the lateral tip position: a) the AFM tip has just made contact with the sample, γ is still 13°; b) z-piezo expands and subsequently γ is reduced; c) more realistic, the cantilever bends under pressure, d) C(V) spectra recorded every 1 μm approaching the sample edge. The curves nearer at the sample edge show a reduced stray capacitance $\overline{C}_{stray}$.

abolished, depending on the required lateral resolution and whether the tip movement was desired or not.

Even if lateral or vertical tip movements due to temperature changes within the AFM do not disturb a particular experiment, uncontrolled high tip pressure could damage the sample as well as the tip itself, or affects the consistency of experiments when pressure dependent samples are used [2]. However, using very low pressure at the begin of an experiment runs the risk of losing tip sample contact at a later time, if unfortunate temperature changes raise the tip from the sample. A way to get around this problems is to use AFM cantilevers with a very low spring constant. The correlation between the force on the AFM tip F_{Hook}, the spring constant c and the displacement l of the tip from the equilibrium position is given by Hooks

---

[2] In the framework of this work, only silicon samples were were investigated that form MOS systems with the dielectric layer and the tip. Both silicon as well as the SiO_2 and ZrO_2 layers are quite hard and mechanically stable materials and no pressure dependence was observed (as long as the oxide layer was not penetrated by the tip). In contrast, GaAs samples do form Schottky contacts with the AFM tip, and show a severe pressure dependence. Furthermore, GaAs samples are soft and the tip can easily penetrate into the sample.

CHAPTER 6. MICROSCOPIC CAPACITANCE SPECTROSCOPY 107

Law:

$$F_{Hook} = c \cdot l \qquad (6.2)$$

$$P = \frac{F_{Hook}}{A} \qquad (6.3)$$

The pressure P between tip and sample depends on the force F_{Hook} and area A of the tip apex. Defining a secure pressure range, a small spring constant c makes it possible to use a larger initial displacement l, which ensures longer tip-sample contact. In addition, by using low c cantilevers, larger displacements due to larger temperature changes do not lead to eventual harmful accumulation of pressure beneath the tip.

Finally, in table 6.3 all the quantitative results of this chapter is summarized. It is intended as a reference for future experiments. The *electrical expansion coefficient* of the z-piezo $\Delta l/\Delta V_{z-piezo}$ was included for its great practical importance during experiments with the particular AFM hardware. For the same reasons the value $\Delta V_{z-piezo}/\Delta DS$ was included, because when using the laser feedback system of the AFM, the force F_{Hook} is adjusted by the desired deflection of the laser signal recorded by the laser detector. If the deflection setpoint DS is adjusted, the feedback system tries to reach the new value by adjusting the voltage on the z-piezo $\Delta V_{z-piezo}$. The value $\Delta V_{z-piezo}/\Delta DS$ gives the change of the deflection setpoint at a certain change of the z-piezo voltage. $\Delta V_{z-piezo}/\Delta DS$ is only useful if the tip is in contact with the sample. It is important to note that the exact value $\Delta V_{z-piezo}/\Delta DS$ depends on the position the laser is reflected from the cantilever. The fluctuation of the laser alignment is considered to be less than 10%, which transforms into a fluctuation of less than 10% for $\Delta V_{z-piezo}/\Delta DS$. The change of the tip-sample-force induced by the different thermal drift scenarios was included for both the common used SCM diamond tips (FMR-CDT) as well as for scanning spreading resistance (SSRM) AFM tips (NCHR-CDT). The spring constants of these types of AFM tips show a great difference. As discussed above, this leads to a corresponding difference in the tip-sample force. The temperature changes ΔT_{sensor} inside the acoustic hood were only measured at a single location behind the camera system. $\Delta l/\Delta T_{sensor}$ and $\Delta F_{Hook}/\Delta T_{sensor}$ were included in table 6.3 as a rough estimate for both the amount of tip drifts Δl in z-direction and the change of tip-sample force ΔF_{Hook} at slow, isothermal temperature changes ΔT_{sensor}. For the same reasons, $\Delta l/\Delta t$ and $\Delta F_{Hook}/\Delta t$ were included to show the reaction to a fast temperature increase due to switching on the feedback laser.

Table 6.3: Quantitative consequences of thermal tip drifts. Positive values again indicate that the AFM tip approaches the sample or the tip sample pressure is increased in case the tip is in contact with the sample. The SSRM tips show a higher force increment at the same amount of thermal expansion due to the much larger spring constant. "isothermal" means that temperature changes occur everywhere at the same time and the same rate. In contrast, "laser" means that some parts of the AFM are heated at different rates, which happens if the feedback laser is switched on (after it was switched off for a long time).

		tip independent
$\Delta l / \Delta T_{sensor}$	(isothermal)	-1 nm/mK
$\Delta l / \Delta t$	(laser)	+30 nm/minute
$\Delta l / \Delta V_{z-piezo}$	(electrical piezo expansion)	14 nm/V

	SCM tip (FMR-CDT)	SSRM tip (NCHR-CDT)
spring constant c	1 N/m	42 N/m
$\Delta V_{z-piezo}/\Delta DS$	8	5
$\Delta F_{Hook}/\Delta T_{sensor}$ (isothermal)	-1 nN/mK	-42 nN/mK
$\Delta F_{Hook}/\Delta t$ (laser)	+30 nN/minute	+1260 nN/minute

6.4 Data Extraction and Analysis

This chapter describes how the relevant data, like the flatband voltage V_{FB} or the oxide-capacitance $\overline{C}_{ox}$ is extracted from the measured C(V) curves. As described in chapter 6.2, all the curves were recorded using an average time of about 6.9s per capacitance value (this corresponds to an average time level of 10). In figure 6.8, again an example of one single C(V) curve is given. The curve is still noisy, therefore, for some experiments the decision was made to record many such curves on one and the same sample spot and average them afterwards to obtain a much smoother curve. It was also shown in chapter 6.2, that performing many C(V) measurements on one sample location does not degrade the sample beneath the tip. Regardless of having only one single C(V) curve or the average of many hundreds of curves, the resulting curve was subjected to further digital smoothing and filtering. Either a *Box* smoothing filter or a *Savitzky-Golay* filter was applied[3]. The difference between the two

---

[3]Box smoothing is similar to a moving average, except that an equal number of points before and after the smoothed value are averaged together with the smoothed value. The

CHAPTER 6. MICROSCOPIC CAPACITANCE SPECTROSCOPY 109

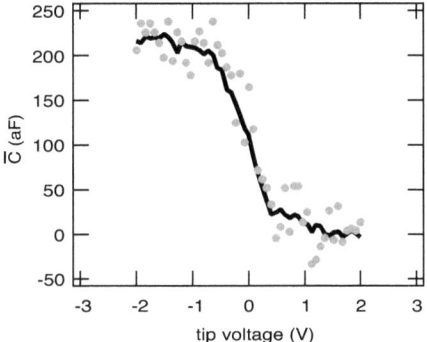

Figure 6.8: Comparison of one single C(V) curve (dots) and the pointwise average of 60 consecutive single C(V) curves on one and the same sample spot (solid line).

smoothing algorithms is the frequency dependence. Savitzky-Golay smoothing better reproduces small scale voltage features of C(V) curves, whereas Box smoothing rather filters them out. The extend to which the curves variability is smoothed out depends on further parameters (number of points that are averaged in Box smoothing, or the order and number of points in Savitzky-Golay smoothing). If different C(V) curves were compared, one and the same smoothing algorithm was used to exclude systematic errors. Both the oxide capacitance as well as the flatband voltage were calculated using smoothed C(V) curves only.

Especially for the comparison of C(V) curves obtained on different sample locations it was important to subtract the orders of magnitude larger stray capacitance $\overline{C}_{stray}$. Common variations of the stray capacitance are 50 fF to 1 pF, depending on the location of the sample where the actual measurement takes place. The position dependence of the stray capacitance can be used as a test for parasitic tip drifts, which is explained in chapter 6.3.2. However, it is also the most severe drawback of QSCS because it prevents the measurement of the oxide capacitance on arbitrarily doped Si samples. It is impossible to calculate the oxide capacitance, if neither the capacitance offset due to the stray capacitance, nor the decrease of the difference between the maximum and minimum capacitance due to the growth of the doping dependent depletion region is known. Therefore the range of possible

---

Savitzky-Golay smoothing filter is a type of Least Squares Polynomial smoothing. It was first proposed by A. Savitzky and M.J.E. Golay in 1964. This description was taken from IgorPro 4.06A help files.

sample doping levels was constrained to the low doped region throughout this work. As explained in chapter 2.3.4 for low doped samples and an ideal C(V) curve without stray capacitance the semiconductor capacitance $\overline{C}_s$ and subsequently the total capacitance $\overline{C}$ is approximately zero. Therefore, the stray capacitance of a realistic C(V) curve on a low doped sample can be identified as the lowest total capacitance value $\overline{C}_{min}$ (equation 2.40), which can then be subtracted from the curve. The difference between the maximum ($\overline{C}_{max}$) and the minimum value of a C(V) curve is the oxide capacitance $\overline{C}_{ox}$ (equation 2.41).

To obtain the flatband voltage V_{FB} of a particular C(V) curve, the flatband capacitance $\overline{C}_{FB} = C_{FB} \times A$ was calculated for an ideal flat capacitor via equations 2.3 and 2.42 and compared to the experimental data. Note that the flatband capacitance strongly depends on the doping level N_A or N_D and the dielectric constant ε_s of the particular Si sample, which was discussed in chapter 2.3.5. Furthermore, it depends on the oxide thickness d and dielectric constant ε_{ox} of the dielectric layer. Therefore, for sample parameters used in this work (in particular: low doped silicon covered with very thin dielectrics or high-ε dielectrics), the normalised flatband capacitance $\overline{C}_{FB}/\overline{C}_{ox}$ ($= C_{FB}/C_{ox}$) is in the range of 0.02 to 0.09 corresponding to a flatband condition occurring at 2 to 9 % of the maximum capacitance of the measured C(V) curve. The flatband voltage V_{FB} can be obtained by taking the voltage of a C(V) curve where the capacitance value has reached the flatband capacitance of the particular sample.

However, the very low values of the normalised flatband capacitance encountered in this work lead to difficulties in obtaining accurate flatband voltage values from noisy C(V) data. The voltage values of a C(V) curve are predetermined and exactly known [4], whereas the capacitance values are measured and show a random uncertainty $\delta\overline{C}$. At a slope S of the C(V) curve the random uncertainty $\delta\overline{C}$ of the capacitance translates into a (random) uncertainty of the voltage δV.

$$\frac{\delta\overline{C}}{\delta V} = S \longrightarrow \delta V = \frac{\delta\overline{C}}{S} \qquad (6.4)$$

Near the lowest capacitance values, the slope S of the C(V) curve is very small, and according to equation 6.4, a small random uncertainty of the capacitance gives a large random uncertainty in the corresponding voltage. Therefore, at low capacitance values, random fluctuation of the measured

---

[4]Every leakage current through the oxide automatically leads to a decrease of the applied tip voltage. However this case would not happen unobserved, because a sudden breakdown of the dielectric layer would be visible via an severe increase of the loss signal measured by the capacitance bridge.

C(V) data lead to a very large *random* error δV in flatband voltage determination.

The simples way to get around this is averaging over a large number of C(V) curves, because the fluctuations are random and will cancel out.

Another possibility is to determine the voltage V_{FB+} at a slightly higher capacitance value $\overline{C}_{FB+} = \overline{C}_{FB} + \Delta \overline{C}$ than the calculated flatband capacitance $\overline{C}_{FB}$. At higher capacitance values the steepness of the C(V) curve is increased which leads to a reduction of the *random* error in the corresponding voltage V_{FB+}. However, now an error ΔV is introduced: $V_{FB+} = V_{FB} + \Delta V$. This error ΔV can be assumed to be *systematic*. This allows to determine flatband voltage *shifts* much more accurate, because the systematic error cancels out during subtractions. To a certain extent, it is also possible to obtain an estimate of ΔV, which can be used to calculate V_{FB}. See chapter 6.6 for more details on this method.

6.5 C(V) Spectroscopy on Macroscopic and Nanoscopic Capacitors

After developing the basic setup of QSCS in chapter 3.2.2, knowing how to treat it and having an idea of its limits from the previous three sections, the next step was to evaluate the quality of the measured data obtained via QSCS. Therefore, both macroscopic measurements on large area 100×100 μm MOS capacitors and AFM based, nanoscopic investigations were carried out on identical pieces of samples, as is schematically shown in figure 6.9. The MOS capacitors' top electrodes were lithographically patterned, sputtered Al. A standard Agilent LCR meter was used to measure the C(V) curves on the MOS capacitors inside a dark box, to prevent the influence of light. For consistency, the nanoscopic capacitance measurements were always performed in close vicinity to the corresponding macroscopic MOS capacitors. The samples were homogeneously p-doped Si-wafers with an acceptor concentration of $N_A = 9.4 \times 10^{14}$ cm^{-3}. Wafers having higher concentration ($N_A = 1 \times 10^{18}$ cm^{-3}) were investigated too and qualitatively showed the same results. Two types of dielectric layers between the tip and the semiconductor were used: industry quality high temperature grown SiO_2 and Metal-Organic Chemical Vapour Deposition (MOCVD)-grown ZrO_2. Prior to the oxidation or deposition process the samples were cleaned using a RCA cleaning process [55] followed by a removal of the native SiO_2 in HF. The SiO_2 layer was created by wet oxidation at 900 °C, and the ZrO_2 layer was deposited at 450 °C in a horizontal hot-wall reactor equipped with a bubbler system for

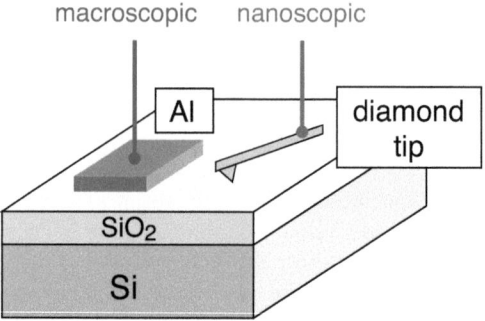

Figure 6.9: Schematic of the experiment. A 100×100 μm MOS capacitor with lithographically patterned Al top electrode was characterised with standard LCR meter equipment. In close vicinity to the Al electrode nanoscopic measurements with a diamond coated AFM-tip connected to an AH2550 capacitance bridge took place.

metal-organic precursor delivery. No post deposition thermal annealing was used for the ZrO_2 covered sample. This results in a rather poor quality ZrO_2 layer. For a more detailed introduction into the oxidation process or MOCVD look at chapter 4.2. The nominal silicon oxide layer thickness was 4 nm and was confirmed by ellipsometric measurements. While SiO_2 has a dielectric constant of $\varepsilon_{ox} = 3.9$, ZrO_2 is a high-ε material with a dielectric constant of approximately $\varepsilon_{ox} \approx 20$. For a direct comparison of the capacitance data, we fabricated dielectric layers with the same dielectric constant to thickness ratio ε_{ox}/d. (SiO_2: $d=$ 4 nm, ZrO_2 : $d=$20 nm). In figure 6.10, a comparison between the C(V) data of a reference MOS capacitor, a simulated C(V) curve and a nanoscopically measured C(V) curve is shown. The nanoscopic curve was obtained by averaging over approximately 80 single C(V) curves to increase the signal to noise ratio. Afterwards, the averaged curve was smoothed and the three orders of magnitude larger stray capacitance was subtracted, as described in chapter 6.4. The shape of all three curves in figure 6.10 is almost identical. According to MOS theory text books [83], C(V) curves are smeared out by interface traps. The slightly smaller slope of the macroscopic reference curve in figure 6.10 (a) compared to the calculated curve (b) can therefore be explained by an interface trap number density N_{it} of about 5×10^{11} cm^{-2}. Although the nanoscopic and the macroscopic C(V) curve were both recorded on nearly the same sample position, the nanoscopic curve is not as steep as the macroscopic curve. There are two possible reasons for that behaviour, which will be assessed now.

- First, a possible reason for this behaviour is the very small area of the AFM tip and an increasing influence of electrostatic edge effects. However, an estimate given in chapter 2.5 of the influence these edge effects have on the shape of measured C(V) curves shows that it is relative implausible that edge effects are the sole reason for the broadening.

- It seems more probable that the broadening is caused by an elevated interface trap density. Whereas the oxide layer of the macroscopic MOS capacitors was protected from the ambient environment by the sputtered top electrode, the AFM based nanoscopic measurement took place at an oxide layer that was constantly exposed to the ambient environment. This leaves the oxide layer vulnerable to contamining situations and agents, which may change the oxide quality.

Maybe, the observed broadening could be explained by a combinations of this two effects. A definitive explanation could not be found in this work and is a subject for future investigations.

The position of the transition between accumulation and depletion on the voltage axis (flatband voltage) can be used to extract various physical information. The position is defined mainly by two factors: the work function of the capacitor's top electrode W_m (if the work function of the semiconductor W_s remains fixed) and the total amount of oxide charge Q_{tot} (refer to equation 2.45). The simulated curve in figure 6.10 (b) shows a flatband voltage of 0 V, because the simulation was performed for a top electrode with the same work function as the semiconductor ($W_{ms} = 0$) and because no oxide charges were included. The flatband capacitance of curve (b) occurs at a normalised value of $\overline{C}/\overline{C}_{ox}$=0.08 . The two measured curves in figure 6.10 (a) and (c) are located at significantly different positions than the calculated curve (b). As described in chapters 2.3.5 and 6.4, the flatband voltages of curve (a) and (c) were obtained by determining the voltages at which the normalised capacitance equals the flatband capacitance of the calculated curve. The difference between the flatband voltages of the two measured curves is $\Delta V_{FB} = 1.6V \pm 0.1V$ corresponding to a work function difference between the two top electrode materials. As the work function for Al is known (W_{Al}=4.2 eV)[109], we can calculate the work function of the diamond tip: $W_{tip}= \Delta V_{FB} + W_{Al} = 5.8eV \pm 0.1eV$.

A closer look to the nanoscopic curve in figure 6.10 (c) reveals that the capacitance values again start to increase after a tip voltage of about +1V. This behaviour is *not* observed in the macroscopic C(V) curve 6.10 (a). One

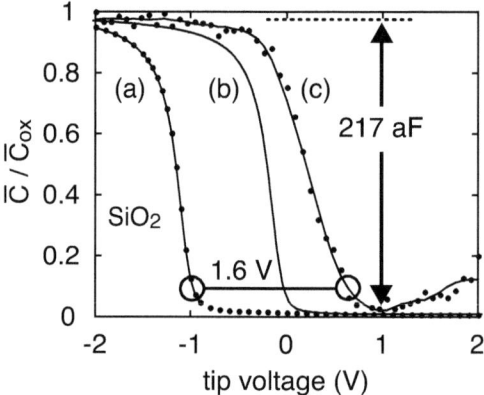

Figure 6.10: (a) C(V) data of a 100×100 μm^2 MOS capacitor with lithographically patterned Al top electrode. (b) Simulated C(V) spectrum of the MOS capacitor assuming no oxide charges and flatband condition at zero bias. (c) C(V) spectrum obtained with a diamond coated AFM-tip connected to an AH2550 capacitance bridge.

can explain the increase of the capacitance in figure 6.10, curve (c) by assuming the movement of mobile ions inside the oxide, which continuously changes the capacitance behaviour when an electric field is applied. However, no mobile ions are present in curve (a), because the top electrode of the macroscopic capacitor leading to curve (a) was created shortly after the oxidation process. One can assume that this somehow shields the underlying SiO_2 from contaminating Na^+- and K^+-ion incorporation. In contrast to that, the oxide area where the nanoscopic measurements via the AFM tip took place were exposed to ambient conditions during the few months prior to the measurements. This is plenty of time to accumulate Na and K, which change the electrical behaviour of oxides. The same effect observed in 6.10 (c) for SiO_2 was also observed for ZrO_2.

To explain what has happened in (c) one has to consider the precise measurement conditions first. The C(V) curve (c) shown in 6.10 was recorded for a voltage ramp starting at a positive tip voltage of +2V and decreasing to -2V. After that, the voltage again jumps to +2V and the next curve was recorded immediately on the same sample spot *without* any delay. Afterwards all curves on one and the same sample spot were used to calculate an average C(V) curve. Curves 1 and 2 in figure 6.11 where recorded on ZrO_2 and SiO_2

respectively, and show a slightly decreasing capacitance for a tip voltage starting at +2V and decreasing to -2V. The time constants for this decrease is below 1 minute.

Two parameters are used to describe the net mobility μ_{ion} of ions inside the oxide ([83] page 442 and [106]): the activation energy E_A which describes how many ions are free to move around and the mobility μ_0 at the limit of total activation of all ions. μ_0 governs the scattering of the drifting ions inside the oxide. The two parameters are combined in

$$\mu_{ion} = \mu_0 \exp(-\frac{E_A}{kT}), \qquad (6.5)$$

where k is the Boltzmann constant and T is the absolute temperature of the oxide. The lower the activation energy E_A and the higher the temperature, the more ions are released to move. Knowing the net mobility μ_{ion} and the applied electric field F_{ox}, one can calculate the distance s the ions move at a certain time t.

$$s = \mu_{ion} F_{ox} t \qquad (6.6)$$

The electric field F_{ox} is determined by the thickness d of the oxide and the applied voltage: V_{tip}: $F_{ox} = V_{tip}/d$. These are the equations needed to calculate the time an ion needs to pass through the dielectric layer. Insertion of $E_A = 0.71$ eV and $\mu_0 = 0.8 \times 10^{-3}$ cm^2/Vs into equation 6.6 leads to a calculated time of $t = 42$s the ions need to pass through the 4nm thick SiO$_2$. This matches well with the data shown in figure 6.11, curve 2. For the 5 times thicker (20nm) ZrO$_2$ sample (curve 1), $E_A = 0.63$ eV and $\mu_0 = 0.7 \times 10^{-3}$ cm^2/Vs results in $t = 58$ s the ions pass through the dielectric. Note, that it is crucial for the calculations to consider that the temperature inside the acoustic hood were the measurements fook place was about 35 °C (308 K). This is significantly higher than the ambient temperature outside the acoustic hood (about 20 °C to 25 °C). Equation 6.5 strongly depends on the temperature T! Values for different ion species found in literature [106]) are E_A=0.66±0.02 eV and μ_0=1.05 cm^2/Vs (within a factor of 2) for Na$^+$-ions and E_A=1.09±0.06 eV and μ_0=0.026 cm^2/Vs (within a factor of 4) for K$^+$-ions. The values found for the samples in this work do not match exactly, but taking the rather coarse estimate into account, they seem quite reasonable. Furthermore, time constants due to a eventually large parasitic stray capacitance and/or resistivity (RC-delay) and the decrease of capacitance due to possible anodic oxidation were investigated, too, but were *not* responsible for the observed effects.

To get rid of the mobile ion induced initial increase of the capacitance, it is possible to insert a delay time of about 2 minutes into the measurement

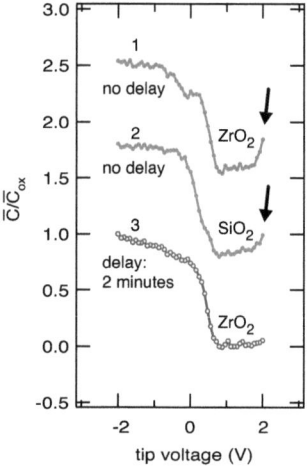

Figure 6.11: Comparison of C(V) spectra recorded without and with a 2 minute delay prior to the measurement. The C(V) curves were first normalised to get rid of any tip area dependence. Arbitrary offsets were added to the curves. Curve 1 and 2 were obtained without any delay on ZrO_2 and SiO_2. Near 2V tip voltage, a higher capacitance was recorded due to mobile ions. The effect is weaker for the much thinner SiO_2. Curve 3 was recorded on ZrO_2 after a 2 minute delay. No effects due to mobile ions is visible.

sequence prior to the actual start of a C(V) spectrum, where the tip voltage is held constant at its start value. Curve 3 in figure 6.11 was recorded this way and shows no initial decrease of the capacitance near +2V due to mobile ions. To avoid inconsistent data, all C(V) curves are recorded using the same voltage sweep direction. From now on all C(V) curves are recorded with an initial delay of 2 minutes.

For future applications it is important to check to what extent the setup is capable of measuring details of the interface trap energy distribution. It was already mentioned that interface traps lead to a decrease of the slope of the C(V) curve between the accumulation and depletion region. If most of the interface traps lie within a small energy interval ($D_{it}(E)$ not constant inside the band gap, equation 2.44), the transition between accumulation and depletion in the C(V) curve contains regions of reduced slope or "kinks". The reason for the interface traps energy distribution and the emergence of kinks

CHAPTER 6. MICROSCOPIC CAPACITANCE SPECTROSCOPY

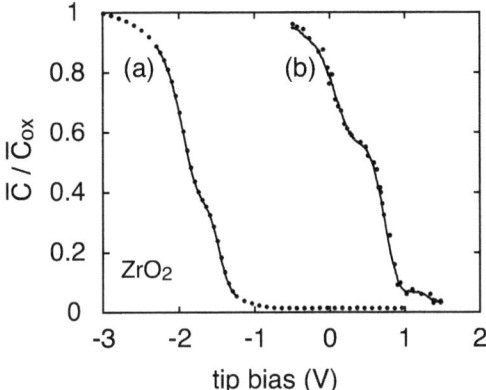

Figure 6.12: Comparison of C(V) data obtained by (a) macroscopic reference measurements on $100\times100\,\mu m^2$ MOS capacitor with lithographically patterned Al and (b) nanoscopic measurements obtained with a diamond coated AFM-tip connected to an AH2550 capacitance bridge. The sample was low doped p-type Si with 20 nm ZrO_2 deposited by MOCVD.

lies within the exact chemical composition and bounding relations of the interface region. To test the ability of our setup to resolve the interface trap energy distribution, we took a rather poor quality ZrO_2 sample. Both the macroscopic reference curve in figure 6.12 (a) as well as the nanoscopic C(V) curve in (b) exhibit a pronounced kink, which demonstrates that both curves have a comparable energetic resolution. The reason for the kink's different position in the reference curve and the nanoscopic curves was not investigated in detail, but is probably correlated to the problem of the different slope of the macroscopic and nanoscopic C(V) curves discussed on pages 112 and 113. Another possibility are statistical variations in the local dielectric properties of the ZrO_2 layer.

Finally, the superior performance of the AH2550 bridge compared to the DI3100 SCM electronics is demonstrated. The AH2550 can be operated under small signal conditions, whereas the SCM obviously operates in the large signal regime in most cases. Furthermore, the DI3100 SCM yields dC/dV data only. The consequences are shown in figure 6.13 where we compared dC/dV data of a ZrO_2 covered sample. The data were measured on the same sample position both with the AH2550 (see figures 6.13 (a)), and with the DI3100-SCM (figure 6.13 (b)). The SCM curve (b) is obviously much

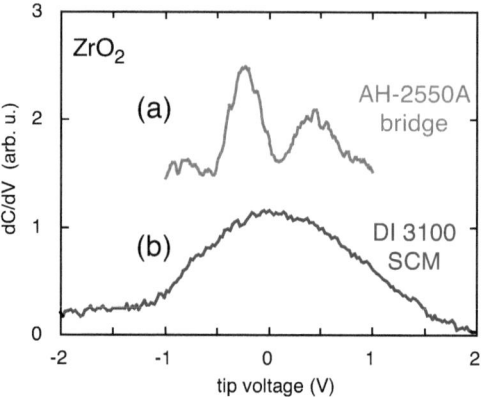

Figure 6.13: Comparison of nanoscopic dC/dV data obtained from the AH2550 capacitance bridge and a standard DI3100 SCM-system. (a) The dC/dV vs. V data of a ZrO_2 coated sample show two narrow peaks which are due to the kink in the ZrO_2 C(V) data. (b) Standard DI3100 SCM measurements on the same sample spot show one broad peak without any features.

broader and has no visible features. In contrast to that, the curve obtained by the AH2550 resolves a double peak behaviour in dC/dV corresponding to the kink in the C(V) curve of figure 6.12.

6.6 Quantitative Investigation of the Properties of a ZrO_2 Layer

In this section, QSCS was applied to investigate the local microscopic properties of a ZrO_2 layer covered Si sample. As sample we used a homogeneously p-doped Si wafer with an acceptor concentration of $N_A = 9.4 \times 10^{14}$ cm^{-3}. The sample dimensions were 2×2 cm^2. Again, the ZrO_2 layer was deposited by Metal Organic Chemical Vapour Deposition (MOCVD) at 450°C. This time, post deposition annealing at 650°C for 5 minutes in forming gas (N_2 + H_2) was used to improve oxide quality. For a more detailed introduction into MOCVD look at chapter 4.2.3. The average thickness of the ZrO_2 layer was determined by ellipsometry and was about 5.12 nm.

Figure 6.14 shows the topography of the ZrO_2 sample recorded a few μm away from the sample edge by a non contact AFM method. The ZrO_2 surface

CHAPTER 6. MICROSCOPIC CAPACITANCE SPECTROSCOPY

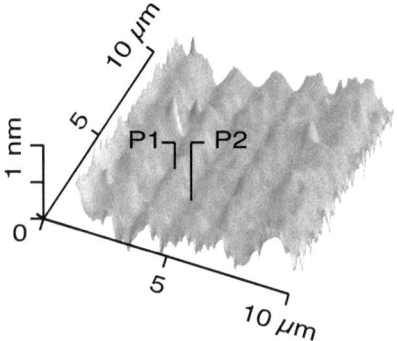

Figure 6.14: Non-contact AFM image showing topographic ripples on the ZrO$_2$ layer close to the edge of the sample. P1 marks a ridge of the ripples, P2 corresponds to the trough.

shows periodic thickness variation (ripples) with an approximate period of about 1700 nm and a height variation of about 0.4 nm. This ripple formation happens near the sample edges probably due to a not ideal precursor flow during the CVD process. Although the deposition parameters that led to the ripple formation was not investigated further, this effect was used to demonstrate the excellent lateral and capacitance resolution of the QSCS-setup. With the QSCS-equipment it is possible to quantitatively analyse the relation between local thickness and the local electrical parameters of the ZrO$_2$ surface shown in figure 6.14.

Figure 6.15 shows a comparison between C(V) curves recorded on the ridge and in the trough of a ripple, corresponding to the markers P1 and P2 in figure 6.14. The nanoscopic C(V) curves were obtained by smoothing and subtracting the stray capacitance from the raw data.

From equation 2.1 one can see the correlation between the oxide capacitance $\overline{C}_{ox}$ and the oxide thickness d, the dielectric constant of the oxide ε_{ox}, and the effective, electrical area of the tip A. As ε_{ox} and A are supposed to be constant, $\overline{C}_{ox}$ is inversely proportional to the thickness d of the oxide. Therefore, a measurement inside a trough (position P2 in figure 6.14) delivers a high maximum capacitance $\overline{C}_{ox,2}$, because the local oxide thickness is reduced. The difference between the maximum capacitance values in figure 6.15 is therefore due to a different oxide thickness d at locations P1 and P2. Another important parameter is the position of the transition between accumulation and depletion conditions, the flatband voltage V_{FB}. As one can see

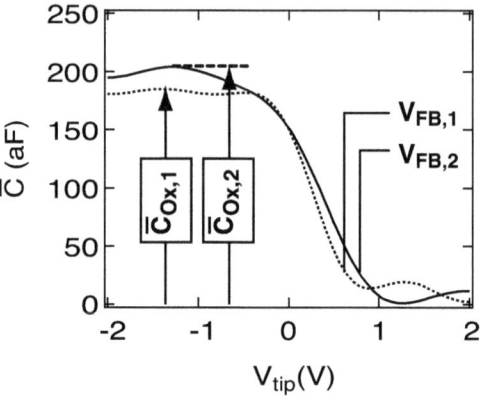

Figure 6.15: Typical C(V) curves taken at P1 and P2 of figure 6.14. $V_{FB,1}$ and $V_{FB,2}$ indicates the corresponding flatband positions. $\overline{C}_{ox,1}$ and $\overline{C}_{ox,2}$ depend on the oxide thickness d at P1 and P2.

in figure 6.15, this parameter changes significantly from a value $V_{FB,1}$ at the sample location P1 to $V_{FB,2}$ at the location P2. The flatband voltage V_{FB} is given by equation 2.45, and depends on the work function difference between the diamond AFM tip and the silicon W_{ms}, the oxide capacitance per area $C_{ox} = \overline{C}_{ox}/A$ and the total amount of oxide charges per area $Q_{tot} = \overline{Q}_{tot}/A$. From the previous chapter 6.5 the work function of the diamond tip is known ($q \cdot W_{tip} = 5.8$ eV). The work function of the silicon depends on the doping level and can be calculated using equations 2.11 and 2.14 ($q \cdot W_s = 4.9$ eV @ $N_A = 9.4 \times 10^{14}$ cm^{-3}). Therefore the work function difference is $q \cdot W_{ms} = 0.9$ eV. As the work function difference is known, the flatband voltage V_{FB} is only determined by the total oxide charge density Q_{tot} inside the oxide and the oxide capacitance C_{ox} (Farad/cm^2), which itself depends only on the oxide thickness d.

Therefore, if the work function difference, the dielectric constant of the oxide and the tip area are known and assumed to be constant throughout the experiment, it is possible to obtain the oxide capacitance and the total oxide charge area density Q_{tot} from C(V) measurements.

The nanoscopic C(V) curves in figure 6.15 were obtained by smoothing the raw data and subtracting the orders of magnitude larger stray capacitance (about 72 fF). The error of determining the flatband voltage due to applying different smoothing algorithms on one and the same C(V) curve

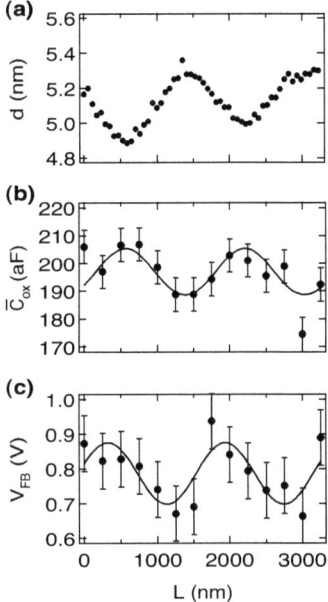

Figure 6.16: (a) : cross section of an arbitrary ZrO$_2$ ripple. The average oxide thickness was measured by ellipsometry and added to the ZrO$_2$ thickness variations determined by AFM. (b): Measured oxide capacitance on different spots along a line L separated by 250 nm. (c): Flatband voltages on the same positions.

turned out to be below 30 mV. This is negligible compared to the deviation of the flatband voltage V_{FB} due to noisy C(V) curves. The impact of noise on the determination of the flatband voltage turned out to be about 80 mV. This was determined by recording different C(V) curves on one and the same sample spot, and determining the flatband voltage as described in chapter 6.4. Afterwards the statistical spread was evaluated. Similarly, noise limited the accuracy of determining the oxide capacitance $\overline{C}_{ox}$ to 6 aF.

Figure 6.16 shows how the local oxide capacitance $\overline{C}_{ox}$ and flatband voltage V_{FB} change on 14 successive locations along a line across a period of the ripples. The distance between the locations was 250 nm. Both the oxide capacitance and flatband voltage were derived from C(V) curves recorded on these sample spots as discussed in 6.4. The flatband capacitance $\overline{C}_{FB}$ at these sample spots is located at only about 2% of the maximum capacitance. As explained in chapter 6.4, in this case it is too inaccurate to determine the flatband voltage directly from the corresponding very low flatband capacitance. Instead, V_{FB+} was determined at 20% of the maximum capacitance. An estimate of the systematic voltage shift ΔV of the measured curve was gained by determining the voltage shift between 2% and 20% of the maximum capacitance of a simulated C(V) curve. (see chapter 6.4). Finally, V_{FB+} and ΔV were used to calculate the flatband voltage values of figure 6.16 (b). The error bars in figure (b) and (c) show deviations of 6 aF and 80 mV respectively, and are due to noise. In contrast to chapter 6.5 only one C(V) curve was recorded per sample spot. Thus, time consuming averaging procedures over many curves could be avoided and the measurement time was reduced to about 12 minutes per sample spot. Therefore, it is still possible to improve the obtained noise level by recording more C(V) curves, even at the expense of longer measurement times. In figure 6.16 (a), a cross section of the ZrO_2 layer topography is shown. The variation of the oxide thickness was measured by non contact AFM. A constant background was added to the height scale using the average thickness of the oxide determined by ellipsometry.

A good approximation of the oxide thickness $d(x)$ at the location x on the sample is

$$d(x) = d_{mean} + d_{mod} \cdot \sin(\alpha + 2\pi \cdot \frac{x}{L_{mod}}) \qquad (6.7)$$

where d_{mean}=5.12 nm is the average thickness of the oxide, L_{mod} is the period of the modulation and α is the phase. The variation of the oxide thickness oscillates around the average thickness, therefore the amplitude d_{mod} is defined as half of the 0.4 nm difference measured by non contact AFM (d_{mod}=0.4nm/2). Figure 6.16 (b) shows the capacitance values $\overline{C}_{ox}$ derived from recorded C(V) curves on the 14 evenly spaced locations. A comparison of figure 6.16 (a) and (b) shows that the periods L_{mod} of the topography and capacitance oscillations are about 1645 nm and in excellent agreement. An increase of the oxide thickness d leads to a decrease of the oxide capacitance $\overline{C}_{ox}$, which is in accordance with equation 2.1. In figure 6.16 (b) a simple sine-fit was applied as a guide for the eye. In figure 6.16 (c), the variation of the flatband voltage V_{FB} is shown. Again a simple sine-fit was applied. According to the fit, the curve is shifted to the left side, and the period of L_{mod}=1613 nm is slightly smaller than the capacitance in figure 6.16 (b). However, this deviations are quite small, and no systematic effect is seen therein. On the whole, the flatband voltage shows the same periodicity as the topography and the capacitance. The capacitance and flatband voltage data *qualitatively* repeat the topography data.

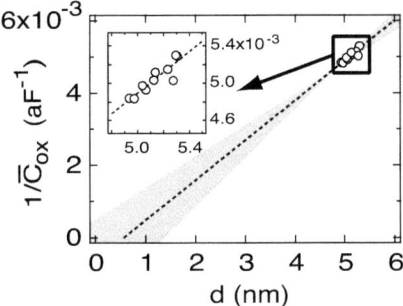

Figure 6.17: The reciprocal of the oxide capacitance ($1/\overline{C}_{ox}$) is plotted versus the thickness of the oxide d on a ZrO$_2$ ripple. The dashed line shows a linear fit to the data, the grey area covers everything within one standard deviation of the uncertainty of the fit. The inset zooms into the data for better clarity.

A verification that there is also a meaningful *quantitative* correlation between topography and oxide capacitance is shown in figure 6.17. Here the reciprocal of the oxide capacitance on the 14 evenly spaced locations is plotted versus the corresponding thickness of the oxide. The oxide thickness $d(x)$ again was the sum of the average thickness from ellipsometry and the variations measured by AFM (equation 6.7). The dotted line fits the measured data which are marked by circles. The grey shaded area corresponds to one standard deviation of the fit parameters. Assuming a constant dielectric constant ε_{ox} throughout the thickness of the oxide, the ideal correlation should be a straight line intersecting the origin of the graph. This requirement is nearly fulfilled by the measured data. The standard deviation is quite small and the fit based on measured data is less than one standard deviation away from the ideal curve that would intersect the graphs origin. The inset shows a magnification of the relevant plot area. From the slope of the curve one can calculate the effective, electrical tip area A, which turned out to be 76×76 nm^2. This result is reasonable for diamond coated AFM tips that, according to datasheets, show a radius of curvature of the tip apex lying in between 100 and 200 nm.

Using equations 2.45 and the capacitance and flatband voltage data, one can calculate the total oxide charge area density Q_{tot}. To compute the areal density, the previously calculated electrical tip area A=76×76 nm^2 was used. The result is shown in figure 6.18. First we find that the oxide charge density was always *positive* on this particular sample. This may be explained by a

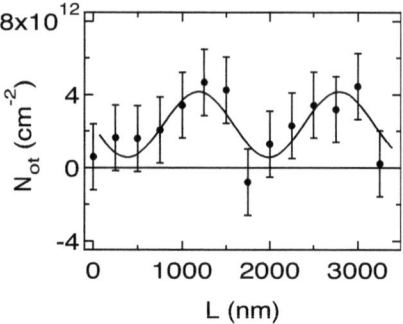

Figure 6.18: Local oxide charge density on a line L of our sample. The total charge number density N_{tot} is largest on the ridge and smallest on the trough of a ripple.

large fraction of fixed oxide charges Q_f that contribute to the total oxide charge, and which are located within the first few nanometres of ZrO_2 layer after the Si-ZrO_2 interface. With about 5 nm the ZrO_2 layer may be too thin for large fractions of eventual negatively charged oxide traps Q_{ot}. The more important result, however, is that the total oxide charge number density N_{tot} ($N_{tot} = Q_{tot}/q$) is not constant and varies between $+1.3 \times 10^{12}$ cm^{-2} at the troughs and $+4.7 \times 10^{12}$ cm^{-2} at the ridges of the modulated sample surface. Due to random uncertainties of 80 mV in determining the flatband voltage, the total oxide charge number density N_{tot} varies by $\pm 1.8 \times 10^{12}$ cm^{-2}. Error bars are set accordingly in figure 6.18. In addition, systematic uncertainties in the work function of the diamond tip of ± 0.1 eV lead to a possible collective shift of the whole curve in figure 6.18 of about $\pm 2 \times 10^{12}$ cm^{-2} (not shown in figure 6.18).

As one can see from figure 6.18, the periodicity of the calculated total charge number density N_{tot} is the same as in the oxide topography. The higher the oxide thickness at a location, the higher is the charge density at that location.

Furthermore, the relative changes of the total oxide charge numbers density N_{tot} in figure 6.18 and oxide thickness in figure 6.16 (a) are far from being equal. The charge density more than triples while the oxide thickness increases only by 0.4 nm (less than 10% of the total thickness). This is an important result which shows that the growth conditions at locations where ripple formation takes place are probably very different from the normal growth mode.

Chapter 7
Critical Remarks and Outlook

Finally, it is time for some critical remarks concerning the work presented here. From a logical point of view, it is not very satisfactory to steadily apply 1D MOS theory on an SPM tip - sample system with a tip diameter of less than 100 nm. In qualitative, conventional SCM this is of no practical relevance, since the large applied voltages introduce deviations obviously much larger than effects due to the small SPM tip. However, the quantitative measurements with the capacitance bridge setup (QSCS) indicate that geometric effects due to the tip shape may already interfere with the measurements. E.g. the reduced slope of the nanoscopically C(V) spectra compared to C(V) spectra of large MOS capacitors could be interpreted in this way. Up to now, various groups [59, 74, 99, 20] performed simulations of the influence of the detailed tip shape on the SCM/capacitance signal, but conventional SCM simply did not provide a high enough spectroscopic resolution to resolve these effects. The required spectroscopic resolution can now be delivered with the presented AH2550A capacitance bridge setup. If it is intended to develop capacitance microscopy into a qualitative tool, it is of paramount importance to obtain also a *quantitative* estimate of the influence of the tip shape. Therefore, in the last months efforts have been made to establish a collaboration with the Institute for Microelectronics to do quantitative, 3D simulation of the tip - sample system. Hopefully, this collaboration will provide quantitative answers to important questions like "How much is the nanoscopic C(V) curve broadened by the impact of the tip shape?" or "Does the flatband voltage shift at the transition from 1D (flat capacitor) to 3D (SPM tip) MOS theory?"

Furthermore, note that especially the determination of the oxide charges in chapter 6.6 was at the limits of the experimental setup's abilities. Especially the thermally induced tip drift due to switching off the feedback laser limited the possible measurement duration and subsequently the amount

CHAPTER 7. CRITICAL REMARKS AND OUTLOOK 126

of measurement data that could be recorded for averaging. In principle, however, it is possible to obtain much more accurate capacitance data and therefore much more accurate oxide charge values. One "only" has to get rid of the tip drift by implementing advanced SPM systems (with closed-loop scanner[1]) and/or innovative approaches to the problem (e.g. implementation of a resistor that dissipates heat to compensate for the cooling induced by switching off the feedback laser). To further increase the accuracy of oxide charge determination, precise values of the tip's work function are needed. Again this is based on an estimate how the flatband voltage is influenced by the shape of the tip.

A simple increase of the measurement accuracy is not the only task for the future. There are some other possible promising projects for future research using the AH2550A capacitance bridge together with an SPM:

- Local, quantitative capacitance measurement on GaAs-InAs-AlGaAs heterostructures are an interesting topic, because the tip-sample system does not show MOS behaviour anymore. Instead, the AFM tip establishes a Schottky contact to the GaAs sample. Depending on the doping type of the GaAs sample, one expects a diode like behaviour. As Schottky junctions are very dependent on the interface properties, the applied pressure between the tip and the sample will have a large impact on the measured capacitance data. Ultimately, the AH2550 bridge setup may allow capacitance spectroscopy on *single* InAs quantum dots. Currently, the first investigations on that topic are already in progress.

- Another promising subject is extending quantitative scanning capacitance microscopy/spectroscopy to the low temperature regime between 4.2K and 300K. One can assume that measurements at low temperatures on InAs quantum dots will lead to better spectroscopic results, because of the suppression of phonon-scattering. It may be possible then to even resolve the atom-like states of the quantum dot via singe dot, low temperature capacitance spectroscopy.

---

[1]In contrast to other scanners, a closed-loop scanner effectively is free of *piezo* drift. The piezo voltage for a certain expansion is dynamically adjusted by aid of a capacitive feedback, that steadily performs measurements of the absolute extension of the piezo. The piezo voltage, therefore, is controlled in a *closed-loop* based on the current piezo position to compensate for piezo creep (and, depending on the design of the closed-loop system, even for other piezo drift scenarios).

Appendix A
List of Symbols

Symbol	Description	Unit
a	linear thermal expansion coefficient	K^{-1}
a_{head}	linear thermal expansion coefficient of the scanning head	K^{-1}
a_{stage}	linear thermal expansion coefficient of the z-stage	K^{-1}
A	area	m^2
A_{FIB}	extent of the damaged regions inside irradiated materials in beam direction	m
B_{FIB}	extent of the damaged regions inside irradiated materials perpenticular to the beam direction	m
c	spring constant of AFM cantilevers	N/m
$\overline{C}, C$	total capacitance of the MOS capacitor	F, F/m^2
$\overline{C}_{adj}$	capacitance of the varactor diode of the SCM module's circuitry	F
$\overline{C}_{FB}, C_{FB}$	total capacitance of the MOS capacitor at the flatband condition	F, F/m^2
$\overline{C}_{FBs}, C_{FBs}$	semiconductor capacitance at the flatband condition	F, F/m^2
$\overline{C}_{FB+}, C_{FB+}$	flatband capacitance with systematic error	F, F/m^2

APPENDIX A. LIST OF SYMBOLS

Symbol	Description	Unit
$\overline{C}_{max}$, C_{max}	capacitance maximum of a C(V) curve	F, F/m^2
$\overline{C}_{min}$, C_{min}	capacitance minimum of a C(V) curve	F, F/m^2
$\overline{C}_{ox}$, C_{ox}	oxide capacitance	F, F/m^2
$\overline{C}_{probe}$	tip-sample capacitance	F
$\overline{C}_s$, C_s	semiconductor capacitance	F, F/m^2
$\overline{C}_{stat}$, C_{stat}	static capacitance	F, F/m^2
$\overline{C}_{stray}$, C_{stray}	stray capacitance	F, F/m^2
$\overline{C}_{stray1}$	stray capacitance	F
$\overline{C}_{stray2}$	stray capacitance	F
$\overline{C}_x$	unknown capacitance	F
$\overline{C}_0$	reference capacitance	F
$\delta\overline{C}$, δC	random capacitance error	F, F/m^2
$\Delta\overline{C}$, ΔC	systematic capacitance error or capacitance difference	F, F/m^2
d	thickness of the dielectric layer	m
D_{it}	interface trap level density	cm^{-2}eV^{-1}
d_{mean}	mean thickness of the dielectric layer	m
d_{mod}	amplitude of dielectric layer thickness variation (ripples)	m
DS	deflection setpoint parameter of the AFM in contact mode	V
$d(x)$	thickness of the dielectric layer at lateral position x	m
E_A	activation energy of mobile ions inside a dielectric layer	eV, J
E_C	conduction band edge	eV, J
E_i	intrinsic Fermi level of the semiconductor	eV, J
E_{Fm}	Fermi level of the gate material	eV, J

APPENDIX A. LIST OF SYMBOLS

Symbol	Description	Unit
E_{Fn}	quasi-Fermi level of electrons in the semiconductor	eV, J
E_{Fs}	Fermi level of the semiconductor material	eV, J
E_g	band gap energy	eV, J
E_V	valence band edge	eV, J
F_{Hook}	force acting an the AFM cantilever if it is bend	N
F	electric field	V/m
F_{ox}	electric field inside the SiO_2 or ZrO_2 layer	V/m
F_s	electric fieled inside the semiconductor	V/m
l	length	m
l_{head}	height of the scanning head of the AFM	m
L_{mod}	lateral period of the dielectric layer thickness variations (ripples)	m
l_{stage}	height of the z-stage of the AFM	m
Δl	vertical displacement of the AFM tip	m
N_A	acceptor concentration	m^{-3}
N_D	donator concentration	m^{-3}
N_f	oxide fixed charges number density	m^{-2}
n_i	intrinsic carrier density of a semiconductor	m^{-3}
N_{it}	interface trapped charge number density	m^{-2}
N_m	mobile ionic charges number density	m^{-2}
N_{ot}	oxide trapped charge number density	m^{-2}
N_{tot}	total oxide charge number density	m^{-2}
$n(x)$	volume charge density of the electrons at a distance x from the semiconductor surface	C/m^3
$\overline{Q}, Q$	charge	C, C/m^2
$\overline{Q}_f, Q_f$	oxide fixed charges density	C, C/m^2
$\overline{Q}_{it}, Q_{it}$	interface trapped charge density	C, C/m^2

APPENDIX A. LIST OF SYMBOLS

Symbol	Description	Unit
$\overline{Q}_m$, Q_m	mobile ionic charge density	C, C/m²
$\overline{Q}_{ot}$, Q_{ot}	oxide trapped charge density	C, C/m²
$\overline{Q}_s$, Q_s	charge density at the semiconductor surface due to free carriers	C, C/m²
$\overline{Q}_{tot}$, Q_{tot}	total oxide charge density	C, C/m²
P	pressure	N/m²
$p(x)$	volume charge density of the holes at a distance x from the semiconductor surface	C/m³
R	dissipative resistance	Ω
R_{SCM}	focussed ion beam radius measured by the impact on the SCM signal	m
R_{Topo}	focussed ion beam radius measured by the impact on the topography	m
$\overline{R}_x$	unknown resistance	Ω
$\overline{R}_0$	reference resistance	Ω
ΔR	radial difference between the lateral extent of topography and SCM signal of FIB irradiated sample spots	m
s	length, distance	m
S	slope of the C(V) curve	F/V
t	time, duration	s
T	temperature	K
T_{sensor}	temperature measured by the sensor behind the camera system of the AFM	K
ΔT	temperature change	K
ΔT_{head}	temperature change of the scanning head	K
ΔT_{stage}	temperature change of the z-stage	K
u_B	dimensionless band bending deep inside the bulk	(1)

APPENDIX A. LIST OF SYMBOLS

Symbol	Description	Unit
u_{Fn}	dimensionless term based on the quasi-Fermi level E_{Fn}	(1)
u_s	dimensionless band bending at the semiconductor surface	(1)
$u(x)$	dimensionless band bending at a distance x from the semiconductor surface	(1)
V_{ACbias}	AC bias voltage applied by the conventional SCM	V
v_B	dimensionless band bending deep inside the bulk	(1)
V_{DCbias}	DC bias, voltage difference between the sample and the AFM tip, $V_{tip} = -V_{DCbias}$	V
V_{excit}	excitation voltage used to measure the differential capacitance	V
V'_{excit}	excitation voltage applied to the high frequency resonator circuitry of the SCM module	V
V_{FB}	flatband voltage	V
V_{FB+}	flatband voltage with systematic error	V
V_G	gate voltage, voltage difference between the gate electrode and the substrate of a MOS capacitor	V
v_s	dimensionless band bending at the semiconductor surface	(1)
V_{tip}	tip voltage, voltage difference between the AFM tip and the sample	V
$v(x)$	dimensionless band bending at a distance x from the semiconductor surface	(1)
δV	random voltage error	V
ΔV	systematic voltage error and voltage difference	V

APPENDIX A. LIST OF SYMBOLS

Symbol	Description	Unit
$\Delta V_{z-piezo}$	voltage change applied to the AFM's z-piezo	V
w	depletion layer width	m
W	diameter of the features on a ZrO_2 covered sample	m
W_{Al}	work function of aluminium	V
W_{tip}	work function of the AFM tip	V
W_m	work function of the gate electrode	V
W_{ms}	work function difference between the gate material or the AFM tip and the semiconductor	V
W_s	work function of the semiconductor	V
x	position	m
Z	impedance	Ω
α	phase	(1)
γ	angle of attachment between the AFM tip and the tip holder	(1)
λ_i	intrinsic Debye length	m
λ_n	extrinsic Debye length for an n-type semiconductor	m
λ_p	extrinsic Debye length for a p-type semiconductor	m
μ_{ion}	ion net mobility inside a dielectric layer	$m^2/(V\,s)$
μ_0	ion mobility inside a dielectric layer at the limit of total activation of all ions	$m^2/(V\,s)$
$\rho(x)$	net volume charge density at a distance x from the semiconductor surface	C/m^3
σ	standard deviation	...
ϕ_B	potential deep inside the bulk	V
ϕ_s	potential at the semiconductor surface	V

APPENDIX A. LIST OF SYMBOLS

Symbol	Description	Unit
$\phi(x)$	potential at a distance x from the semiconductor surface	V
χ	electron affinity of the semiconductor	V
ψ_B	band bending deep inside the bulk	V
ψ_s	band bending on the semiconductor surface	V
$\psi(x)$	band bending at a distance x from the semiconductor surface	V

Appendix B
List of Constants

Symbol	Description	Value
k	Boltzmann constant	$1.3806503(24) \times 10^{-23}$ J K^{-1}
q	charge of the electron	$1.602176462(63) \times 10^{-19}$ C
ε_{ox}	dielectric constant of the dielectric layer	SiO$_2$: 3.9, ZrO$_2$: 20
ε_s	dielectric constant of the semiconductor	Si: 11.9
ε_0	permittivity of free space	$8.854187817 \times 10^{-12}$ F m^{-1}

Appendix C
List of Publications

Publications in Context of this Thesis
Publications in Peer-Reviewed Journals

- W. Brezna, S. Harasek, E. Bertagnolli, E. Gornik, J. Smoliner, H. Enichlmair; "Scanning capacitance microscopy with ZrO2 as dielectric material", Journal of Applied Physics, Vol. 92(5), 2002, pp. 2144-2148

- W. Brezna, H. Wanzenboeck, A. Lugstein, E. Bertagnolli, E. Gornik, J. Smoliner; "Focussed ion beam induced damage in silicon studied by scanning capacitance microscopy", Semiconductor Science and Technology 18, 2003, pp. 195-198

- W. Brezna, H. Wanzenboeck, A. Lugstein, E. Bertagnolli, E. Gornik, J. Smoliner, "Scanning Capacitance Microscopy Investigations of Focused Ion Beam Damage in Silicon", Physica E, Vol. 19, 2003, pp. 178-182

- W.Brezna, M.Schramboeck, A.Lugstein, S.Harasek, H.Enichlmair, E. Bertagnolli, E.Gornik, J.Smoliner; "Quantitative Scanning Capacitance Spectroscopy", Applied Physics Letters, Vol. 83(20), 2003, pp. 4253-4255

- W.Brezna, S.Harasek, A.Lugstein, T.Leitner, H.Hoffmann, E. Bertagnolli, J.Smoliner; "Mapping of local oxide properties by quantitative scanning capacitance spectroscopy", Journal of Applied Physics, Vol. 97, 2005, pp. 093701-1 – 093701-4

- W.Brezna, T.Roch, G.Strasser, J.Smoliner; "Quantitative Scanning Capacitance Spectroscopy on GaAs and InAs Quantum Dots" submitted to Semiconductor Science and Technology, 2004

Conference Proceedings

- W. Brezna, S. Harasek, H. Enichlmair, E. Bertagnolli, E. Gornik, J. Smoliner; "ZrO2 as Dielectric Material for Device Characterization with Scanning Capacitance Microscopy", Electrochemical Society Proceedings, Vol. 3, 2003, pp. 378–385

- S. Harasek, H.D. Wanzenboeck, W. Brezna, J. Smoliner, E. Gornik, and E. Bertagnolli; "Utilizing MOCVD for high-quality Zirconium Dioxide Gate Dielectrics in Microelectronics", Electrochemical Society Proceedings, Vol. 8, 2003, pp. 894–899

- W.Brezna, B.Basnar, S.Golka, H.Enichlmair, J.Smoliner; "Calibrated Scanning Capacitance Microscopy for Two-Dimensional Carrier Mapping of n-type Implants in p-doped Si-Wafers", 27th International Conference on the Physics of Semiconductors ICPS-27 Proceedings (American Institute of Physics Proceedings), to be published

Contributed Talks

- W. Brezna, H. Wanzenboeck, A. Lugstein, E. Bertagnolli, E. Gornik, J. Smoliner, "Focused Ion Beam induced Damage in Silicon investigated with Scanning Capacitance Microscopy", 4th International Symposium on Nanostructures and Mesoscopic Systems NANOMES-4, 2003, Tempe, Arizona

- W. Brezna, S. Harasek, H. Enichlmair, E. Bertagnolli, E. Gornik, J. Smoliner; "ZrO2 as Dielectric Material for Device Characterization with Scanning Capacitance Microscopy", 203rd Meeting of the Electrochemical Society, 2003, Paris, France

- S. Harasek, H.D. Wanzenboeck, W. Brezna, J. Smoliner, E. Gornik, E. Bertagnolli; "Utilizing MOCVD for high-quality zirconium dioxide gate dielectrics in microelectronics"; 203rd Meeting of the Electrochemical Society, 2003, Paris, France,

- W. Brezna, M. Schramboeck, A. Lugstein, S. Harasek, H. Enichlmair, D. Rakoczy, E. Bertagnolli, E. Gornik, J. Smoliner; "Quantitative Scanning Capacitance Spectroscopy", 8th International Conference on Nanometer-Scale Science and Technology NANO-8, 2004, Venice, Italy

Invited Talks

- W. Brezna, M. Schramboeck, S. Harasek, A, Lugstein, H. Enichlmair, E. Bertagnolli, E. Gornik, J. Smoliner; "Quantitative Scanning Capacitance Spectroscopy", 27th International Conference on the Physics of Semiconductors ICPS-27, 2004, Flagstaff, Arizona

Poster Presentations

- W.Brezna, H.Wanzenböck, A.Lugstein, E.Bertagnolli, E.Gornik, J.Smoliner; "Focused Ion Beam induced Damage in Silicon investigated with Scanning Capacitance Microscopy", Gesellschaft für Mikroelektronik (GMe) Forum, 2003, Vienna, Austria

- W.Brezna, T.Roch, G.Strasser, J.Smoliner; "Quantitative Scanning Capacitance Spectroscopy on GaAs and InAs Quantum Dots", Gesellschaft für Mikroelektronik (GMe) Forum, 2005, Vienna, Austria

Co-Authorship in other Publications

Publications in Peer-Reviewed Journals

- A. Lugstein, W. Brezna, M. Stockinger, B. Goebel, L. Palmetshofer, E. Bertagnolli, "Nonuniform-channel MOS device", Applied Physics A, Vol. 76(7), 2003, pp. 1035–1039

- A. Lugstein, W. Brezna, G. Hobler, E. Bertagnolli; "Method to characterize the three-dimensional distribution of focused ion beam induced damage in silicon after 50 keV Ga^+ irradiation", Journal of Vacuum Science and Technology A, Vol. 21(5), 2003, pp. 1644–1648

- G.Fasching, K.Unterrainer, W.Brezna, J.Smoliner, G.Strasser; "Tracing deeply buried InAs/GaAs quantum dots using atomic force microscopy and wet chemical etching", Applied Physics Letters, Vol. 86(6), 2005, pp. 063111-1 – 063111-3

Conference Proceedings

- A. Lugstein, W. Brezna, E. Bertagnolli; "Impact of focused ion beam assisted front end processing on n-MOSFET degradation", 40th IEEE International Reliability Physics Symposium IRPS Proceedings, 2002, pp. 369–375

- A. Lugstein, W. Brezna, B. Goebel, L. Palmetshofer, E. Bertagnolli; "Post-Process CMOS Front End Engineering With Focused Ion Beams", 32nd European Solid-State Device Research Conference ESSDERC Proceedings, 2002, pp. 111–114

- H.Wanzenboeck, S.Harasek, W.Brezna, A.Lugstein, H.Langfischer, E. Bertagnolli, U.Grabner, G.Hammer, P.Pongratz; "FIB-TEM characterization of locally restricted implantation damage", Materials Research Society Symposium Proceedings, Vol. 738, 2003, pp. 57–62

- G.Hobler, A.Lugstein, W.Brezna, E.Bertagnolli; "Simulation of focused ion beam induced damage formation in crystalline silicon", Materials Research Society Symposium Proceedings, Vol. 792, 2004, pp. 635–640

- H.Wanzenboeck, S.Harasek, H.Langfischer, B.Basnar, W.Brezna, J. Smoliner, E.Bertagnolli; "Local modification of microstructure and of properties by FIB-CVD", Materials Research Society Symposium Proceedings, Vol. 792, 2004, pp. 453–457

Contributed Talks

- A. Lugstein, W. Brezna, E. Bertagnolli, "Impact of focused ion beam assisted front end processing on n-MOSFET degradation" 40th IEEE International Reliability Physics Symposium (IRPS), Dallas, 2002

- A. Lugstein, W. Brezna, B. Goebel, L. Palmetshofer, E. Bertagnolli, "Post-Process CMOS Front End Engineering With Focused Ion Beams", 32nd European Solid-State Device Research Conference ESSDERC, 2002, Florence, Italy

Poster Presentations

- H. Wanzenböck, S. Gergov, W. Brezna, E. Bertagnolli; "Local deposition of silicon oxide for phase shift photomasks", Informationstagung Mikroelektronik (ME), 2001, Vienna, Austria

- H. D. Wanzenböck, S. Harasek, H. Langfischer, W. Brezna, J. Smoliner, E. Bertagnolli; "Deposition Mechanism of oxide thin films manufactured by a focused energetic beam process", Materials Research Society Fall Meeting (MRS), 2002, Boston, Massachusets

Bibliography

[1] *Andeen-Hagerling AH2550A 1kHz Ultra-Precision Capacitance Bridge.* Manual.

[2] *Datasheet for CDT-FMR AFM Probes from NanoWorld,.* http://nanoworld.com/.

[3] Scanning Capacitance Microscopy (SCM), Support Note No. 289, Rev. A. Technical report, Digital Instruments Veeco Metrology, 2000.

[4] E. Bertagnolli A. Lugstein, W. Brezna. In *IEEE Internat. Reliability Phys. Symp. IRPS proceedings, pp. 369-375*, 2002.

[5] J. Akila and S. S. Wadhwa. *Rev. Sci. Instr.*, 66(3):2517–19, 1995.

[6] D. Alvarez, M. Fouchier, J. Kretz, J. Hartwich, S. Schoenmann, and W. Vandervorst. *Microelectron. Eng.*, 73/74:910–915, 2004.

[7] G. B. Assayag, C. Vieu, J. Gierak, P. Sudraud, and A. Corbin. *J. Vac. Sci. Technol. B*, 11:2420, 1993.

[8] G. Baccarani and M. Severi. *IEEE Transact. Electron Devices*, ED21:122, 1974.

[9] M. Balog, M. Schieber, M. Michman, and S. Patai. *Thin Solid Films*, 47:109, 1977.

[10] M. Balog, M. Schieber, M. Michman, and S. Patai. *J. Electrochem. Soc.*, 126:1203, 1979.

[11] B. Basnar, G. Friedbacher, H. Brunner, T. Vallant, U. Mayer, and H. Hoffmann. *Appl. Surf. Sci.*, 171(3-4):213, 2001.

[12] B. Basnar, S. Golka, E. Gornik, S. Harasek, E. Bertagnolli, M. Schatzmayr, and J. Smoliner. *J. Vac. Sci. Technol. B*, 19(5):1808–1812, 2001.

BIBLIOGRAPHY 141

[13] Gert Binnig, C. F. Quate, and C. Gerber. *Phys. Rev. Lett.*, 56(9):930, 1986.

[14] Gert Binnig and Heinrich Rohrer, 1981. European Patent EP27517.

[15] Axel Born. *Nanotechnologische Anwendungen der Rasterkapazitätsmikroskopie und verwandter Rastersondenmethoden.* PhD thesis, Universität Hamburg, 2000.

[16] O. Bowallius and S. Anand. *Material Sci. Semicond. Processing*, 4:81–84, 2001.

[17] W. Brezna, T. Roch, G. Strasser, and J. Smoliner. *submitted to Semicond. Sci. Technol.*

[18] D. D. Bugg and P. J. King. *J. Phys. E*, 21:147–151, 1988.

[19] G. H. Buh, H. J. Chung, C. K. Kim, J. H. Yi, I. T. Yoon, and Y. Kuk. *Appl. Phys. Lett.*, 77:106, 2000.

[20] G. H. Buh, J. J. Kopanski, J. F. Marchiando, A. G. Birdwell, and Y. Kuk. *J. Appl. Phys.*, 94(4):2680–2685, 2003.

[21] W. K. Chim, K. M. Wong, Y. T. Yeow, Y. D. Hong, Y. Lei, L. W. Teo, and W. K. Choi. *IEEE Electr. Dev. Lett.*, 24(10):667–670, 2003.

[22] D. G. Colombo, D. C. Gilmer, J. V. G. Young, S. A. Campbell, and W. L. Gladfelter. *Chem. Vap. Deposition*, 4:220, 1998.

[23] J. S. Custer, M. O. Thompson, D. C. Jacobson, J. M. Poate, S. Roorda, W. C. Sinke, and F. Spaepen. *Appl. Phys. Lett.*, 64:437, 1994.

[24] C.Y.Nakagura, D.L.Heterington, M.R.Shaneyfelt, P.J.Shea, and A.N.Erickson. *Appl. Phys. Lett.*, 75:2319, 1999.

[25] C.Y.Nakakura, P.Tanyunyong, D.L.Hetherington, and M.R.Shaneyfelt. *Rev. Sci. Instr.*, 74:127, 2003.

[26] P. DeWolf, R. Stephenson, T. Trenkler, T. Clarysse, T. Hantschel, and W. Vandervorst. *J. Vac. Sci. Technol. B*, 18(1):361–368, 2000.

[27] D.Goghero, V.Ranieri, and F.Gazzano. *Appl. Phys. Lett.*, 81:1824, 2002.

[28] O. Douheret, S. Arnand, C. A. Barrios, and S. Lourdudoss. *Appl. Phys. Lett.*, 81:960, 2002.

[29] H. Edwards, V. A. Ukraintsev, R. San Martin, F. S. Johnson, P. Menzand S. Walsh, S. Ashburn, K. S. Wills, K. Harvey, and M.-C. Chang. *J. Appl. Phys.*, 87(3):1485–1495, 2000.

[30] Pierre Eyben. *Scanning spreading resistance microscopy: High resolution two-dimensional carrier profiling of semiconductor structures*. PhD thesis, Katholieke Universiteit Leuven, 2004.

[31] F.Gianazzo, D.Goghero, V.Ranieri, S.Mirabella, and F.Priolo. *Appl. Phys. Lett.*, 83:2659, 2003.

[32] F.Gianazzo, F.Priolo, V.Ranieri, and V.Privitera. *Phys. Lett.*, 7:2565, 2000.

[33] F. Giannazzo, F. Priolo, V. Raineri, and V. Privitera. *Appl. Phys. Lett.*, 76(18):2565, 2000.

[34] Sebastian Golka. Rasterkapazitätsmikroskopie an Dotierprofilen in Halbleitern. Master's thesis, Institut für Festkörperelektronik, Technische Universität Wien, 2001.

[35] A. Grafov, E. Mazurenko, G. A. Battiston, and P. Zanella. *Journ. de Physique*, IV(5):497–502, 1995.

[36] H. G. Hansma. *J. Vac. Sci. Technol. B*, 14(2):1390, 1996.

[37] S. Harasek. *Zirkoniumdioxiddünnfilme als hoch-ε Gateisolatoren für die Siliziumtechnologie*. PhD thesis, Institut für Festkörperelektronik, Technische Universität Wien, 2003.

[38] S. Harasek, H.Wanzenboeck, and E. Bertagnolli. *J. Vac. Sci. Technol. A*, 1(2):653–659, 2003.

[39] U. Heider and O. Weis. *Rev. Sci. Instr.*, 64(12):3534, 1993.

[40] O. Heil, 1935. British Patent 439457.

[41] J. W. Hong, S. M. Shin, C. J. Kang, Y. Kuk, Z. G. Khim, and Sang il Park. *Appl. Phys. Lett.*, 75(12):1760–1762, 1999.

[42] H.Yamamoto, T.Takahashi, and I.Kamiya. *Appl. Phys. Lett.*, 77:1994, 2000.

[43] T. Ishitani and T. Yaguchi. *Microscopy Research and Technique*, 35:320, 1996.

[44] F. Campabadal J. M. Rafi. *Microelectronics Reliability*, 40:1567–1572, 2000.

[45] J. Blanc J. R. Matey. *J. Appl. Phys.*, 57(5):1437–1444, 1985.

[46] H. Janocha and K. Kuhnen. *Sensors and Actuators*, A79(2):83–89, 2000.

[47] J.F.Marchiando, J.J.Kopanski, and J.R.Lowney. *J. Vac. Sci. Technol. B*, 16:463, 1998.

[48] J.J.Kopanski, J.F.Marchiando, and B.G.Rennex. *J. Vac. Sci. Technol. B*, 18:409, 2000.

[49] J.R.Lowney J.J.Kopanski, J.F.Marchiando. *J. Vac. Sci. Technol. B*, 14:242, 1996.

[50] J.K.Leong, J.McMurray, and C.C.Wiliams. *J. Vac. Sci. Technol. B*, 14:3113, 1996.

[51] J.S.McMurray, J. Kim, and C.C. Williams. *J. Vac. Sci. Technol. B*, 15(4):1011, 1997.

[52] H. Jung and D.-G. Gweon. *Rev. Sci. Instr.*, 71(4):1896–1900, 2000.

[53] D. Kahng and M. M. Atalla. *IRE Solid-State Device Res. Conf., Carnegie Institute of Technology*.

[54] C. J. Kan, C. K. Kim, J. D. Lera, Y. Kuk, K. M. Mang, J. G. Lee, K. S. Suh, and C. C. Williams. *Appl. Phys. Lett.*, 71:1546, 1997.

[55] W. Kern and D. A. Poutinen. *RCA Rev.*, 31:187, 1970.

[56] C. K. Kim, I. T. Yoon, Y. Kuk, and H. Lim. *Appl. Phys. Lett.*, 78(5):613–615, 2001.

[57] R. N. Kleinman, M. L. O'Mally, F. H. Baumann, J. P. Garno, and G. L. Timp. *J. Vac. Sci. Technol. B*, 18:2038, 2000.

[58] K. Kobashi, H. Yamada, and K. Matsushige. *Appl. Phys. Lett.*, 81:2629, 2002.

[59] S. Lanyi, J. Török, and P. Rehurek. *J. Vac. Sci. Technol. B*, 14(2):892–896, 1996.

[60] L.Ciampolini, M.Ciappa, P.Malberti, W.Fichtner, and V.Ranieri. *(preprint, submitted)*.

[61] D. T. Lee, J. P. Pelz, and B. Bhushan. *Rev. Sci. Instr.*, 73:3525, 2002.

[62] C. Lehrer, L. Frey, S. Petersen, and H. Ryssel. *J. Vac. Sci. Technol. B*, 19:2533, 2001.

[63] C. Lehrer, L. Frey, S. Petersen, Th. Sulzbach, O. Ohlsson, Th. Dziomba, H. U. Danzebrink, and H. Ryssel. *Microelectron. Eng.*, 57/58:721, 2001.

[64] S. Liang and L. J. Qiao adn W. Y. Chu. *Mat. Lett.*, 57(5-6):1135, 2003.

[65] J.E. Lilienfeld, 1930. U.S. Patent 1745175.

[66] H.-N. Lin, C. Sy-Hann, L. Yuh-Zheng, and C.Show-An. *J. Vac. Sci. Technol. B*, 19(1):308, 2001.

[67] Kun Liu, Bo Zhang, Mingfang Wan, J.H. Chu, C. Johnston, and S. Roth. *Appl. Phys. Lett.*, 70(21):2891, 1997.

[68] M. Ludwig, M. Rief, L. Schmidt, H. Li, F. Oesterhelt, M. Gautel, and H. E. Gaub. *Appl. Phys. A (Mat. Sci. Proc.)*, A68(2):173, 1999.

[69] A. Lugstein, B. Basnar, G. Hobler, and E. Bertagnolli. *J. Appl. Phys.*, 92(7):4037–4042, 2002.

[70] Y. Ma, Y. Ono, L. Stecker, D. R. Evans, and S. T. Hsu. *Tech. Dig. Int. Electron Dev. Meeting*, 149, 1999.

[71] A. Majumdar, J. Lai, M. Chandrachood, O. Nakabeppu, Y. Wu, and Z. Shi. *Rev. Sci. Instr.*, 66(6):3584, 1995.

[72] K. M. Mang, Y. Kuk, J. Kwon, Y. S. Kim, D. Jeon, and C. J. Kang. *Europhy. Lett.*, 67(2):261–266, 2004.

[73] S. Matsui and Y. Ochiai. *Nanotechnology*, 7:247–258, 1996.

[74] J. S. McMurray and C. C. Williams. In *AIP Conference Proceedings, 449(Characterization and Metrology for ULSI Technology)*, pp. 731–735, 1998.

[75] J. Melngailis. *J. Vac. Sci. Technol. B*, 5:469, 1987.

[76] J. Melngailis. In *Proceedings of SPIE-The International Society for Optical Engineering, Electron-Beam, X-Ray and Ion-Beam Submicrometer Lithographies for Manufacturing 1465, pp. 36–49*, 1991.

[77] J. Melngailis, C. R. Musil, E. H. Stevens, E. M. Kellog M. Utlaut, R. T. Post, M. W. Geis, and R. W. Mountain. *J. Vac. Sci. Technol. B*, 4:176, 1986.

[78] R. Menzel, K. Gärtner, W. Wesch, and H. Hobert. *J. Appl. Phys.*, 88:5658, 2000.

[79] J. L. Moll. *Institute of Radio Engineers Wescon Convention Record*, Part 3:32, 1959.

[80] S. Namba. *Nuc. Inst. and Meth. in Phys. Research*, B39:504–510, 1989.

[81] G. Neubauer, A. Erickson, C. C. Williams, M. Rodgers, and D. Adderton. *J. Vac. Sci. Technol. B*, 14:426, 1996.

[82] T. Ngai, W. J. Qi, R. Sharma, J. Fretwell, X. Chen, J. C. Lee, and S. Banerjee. *Appl. Phys. Lett.*, 76:502, 2000.

[83] E.H. Nicollian and J.R. Brews. *MOS (Metal Oxide Semiconductor) Physics and Technology*. Wiley-Interscience, 1982.

[84] J. N. Nxumalo, T. Tran, Y. Li, and D. J. Thomson. In *Proc. IEEE 37th Annual Int. Reliability Physics Symposium*, 1999.

[85] A. Olbrich, B. Ebersberger, C. Boit, Ph. Niedermann, W. Hänni, J.Vancea, and H. Hoffmann. *J. Vac. Sci. Technol. B*, 17:1570, 1999.

[86] M. L. O'Malley, G. L. Timp, S. V. Moccio, J. P. Garno, and R. N. Kleiman. *Appl. Phys. Lett.*, 74(2):272–274, 1999.

[87] M. L. O'Malley, G. L. Timp, S. V. Moccio, J. P. Garno, and R. N. Kleiman. *Appl. Phys. Lett.*, 74(24):3672–3674, 1999.

[88] R. C. Palmer, E. J. Denlinger, and H. Kawamoto. *RCA Review*, 43:194–211, 1982.

[89] W. G. Pfann and C. G. B. Garrett. *Proc. Inst. Radio Eng.*, 47:2011, 1959.

[90] H. O. Pierson. *Handbook of Chemical Vapor Deposition*. Noyes Publications, New York, 1999.

[91] K. C. Popat, S. Sharma, R. W. Johnson, and T. A. Desail. *Surf. and Interf. Anal.*, 35(2):205, 2003.

[92] C. Powell, J. Oxley, and J. Blocher. *Vapor Deposition*. John Wiley & Sons, 1966.

[93] P. D. Prewett and G. L. R. Mair. *Fokused Ion Beams from Liquid Metal Ion Sources*. Research Studies Press, 1991.

[94] C. A. J. Putman, K. O. van der Werf, B. G. de Grooth, and N. F. van Hulst. *Appl. Phys. Lett.*, 64(18):2454, 1994.

[95] S. Richter, M. Geva, J. P. Garno, and R. N. Kleinman. *Appl. Phys. Lett.*, 77:456, 2000.

[96] P. D. Rose, S. J. Brown, G. A. C. Jones, and D. A. Ritchie. *Microelectron. Eng.*, 41/42:229, 1998.

[97] R. S. Rosler, W. C. Benzing, and J. Baldo. *Solid State Technology*, 19(6):45–50, 1976.

[98] R.Stephenson, A.Verhulst, P.DeWolf, M.Caymax, and W.Vandervorst. *J. Vac. Sci. Technol. B*, 18:405, 2000.

[99] D. M. Schaadt and E. T. Yu. *J. Vac. Sci. Technol. B*, 20(4):1671–1676, 2002.

[100] J. Schmidt, D.H. Paoport, G. Behme, and H.J.Fröhlich. *J. Appl. Phys.*, 86(12):7094, 1999.

[101] G. M. Shedd, H. Lezec, A. D. Dubner, and J. Melngailis. *Appl. Phys. Lett.*, 49(23):1584, 1986.

[102] C. C. Shen, J. Murguia, N. Goldsman, M. Peckerar, J. Melngailis, and D. A. Antoniadis. *IEEE Transactions on Electron Devices*, 45:453, 1998.

[103] S. Shin, J.-I. Kye, U. H. Pi, Z. G. Khim, J. W. Hong, Sang il Park, and S. Yoon. *J. Vac. Sci. Technol. B*, 18(6):2664–2668, 2000.

[104] R. C. Smith, T. Ma, N. Hoilien, L. Y. Tsung, M. J. Bevan, L. Colombo, J. Roberts, S. A. Campbell, and W. L. Gladfelter. *Adv. Mater. Opt. Electron.*, 10:105, 2000.

[105] J. Smoliner, B. Basnar, S. Golka, E. Gornik, B. Löffler, M. Schatzmayr, and H. Enichlmair. *Appl. Phys. Lett.*, 79(19):3182–3184, 2001.

[106] J. P. Stagg. *Appl. Phys. Lett.*, 31(8):532–533, 1977.

[107] R. Stephenson, P. DeWolf, T. Trenkler, T. Hantschel, T. Clarysse, P. Jansen, and W. Vandervorst. *J. Vac. Sci. Technol. B*, 18(1):555, 2000.

[108] S. Sundararajan and B. Bhushan. *Sensors and Actuators A (Physical)*, A101(3):338–351, 2002.

[109] S.M. Sze. *Physics of Semiconductor Devices*. Wiley-Interscience, 2nd edition, 1981.

[110] V. A. Ukraintsev, F.R. Potts, R.M. Wallace, L.K. Magel, H. Edwards, and M.-C. Chang. In *AIP Conf. Proc. 449*, 1998.

[111] V.Ranieri and S.Lombardo. *J. Vac. Sci. Technol. B*, 18:545, 2000.

[112] A. Wadas. *Journ. of Magnetism and Magnetic Mat.*, 78(2):263, 1989.

[113] R. Wigren and R. Erlandsson. *Rev. Sci. Instr.*, 67(1):322–4, 1996.

[114] T. Winzell, S. Anand, I. Maximov, E. L. Sarwe, M. Graczyk, and H. J. Whitlow L. Montelius. *Nucl. Instrum. Methods*, B173(4):447–454, 2001.

[115] A. Yamaguchi and T. Nishikawa. *J. Vac. Sci. Technol. B*, 13:962, 1995.

[116] L. Yanhong, J. Yong, J. Xigao, L.Lin, L. Yuguo, and C. Chiming. *Chin. Sci. Bull.*, 47(21):1761, 2002.

[117] Y.Huang, C.C.Williams, and J.Slinkman. *Appl. Phys. Lett.*, 66:344, 1995.

[118] R. J. Young and J. Puretz. *J. Vac. Sci. Technol. B*, 13:2576, 1995.

[119] V. V. Zavyalov, J. S. McMurray, S. D. Stirling, C. C. Williams, and H. Smith. *J. Vac. Sci. Technol. B*, 18(1):549, 2000.

[120] V. V. Zavyalov, J. S. McMurray, and C. C. Williams. *Rev. Sci. Instr.*, 70(1):158–164, 1999.

[121] V. V. Zavyalov, J. S. McMurray, and C. C. Williams. *J. Appl. Phys.*, 85(11):7774–7783, 1999.

Die VDM Verlagsservicegesellschaft sucht für wissenschaftliche Verlage abgeschlossene und herausragende

Dissertationen, Habilitationen, Diplomarbeiten, Master Theses, Magisterarbeiten usw.

für die kostenlose Publikation als Fachbuch.

Sie verfügen über eine Arbeit, die hohen inhaltlichen und formalen Ansprüchen genügt, und haben Interesse an einer honorarvergüteten Publikation?

Dann senden Sie bitte erste Informationen über sich und Ihre Arbeit per Email an *info@vdm-vsg.de*.

Sie erhalten kurzfristig unser Feedback!

VDM Verlagsservicegesellschaft mbH
Dudweiler Landstr. 99   Telefon +49 681 3720 174
D - 66123 Saarbrücken   Fax   +49 681 3720 1749
www.vdm-vsg.de

Die VDM Verlagsservicegesellschaft mbH vertritt

Printed by Books on Demand GmbH, Norderstedt / Germany